STATISTICAL ECOLOGY

STATISTICAL ECOLOGY
A PRIMER ON METHODS
AND COMPUTING

John A. Ludwig
CSIRO Division of Wildlife and Ecology
Deniliquin, NSW, Australia

James F. Reynolds
San Diego State University
San Diego, California

A WILEY-INTERSCIENCE PUBLICATION
JOHN WILEY & SONS
NEW YORK • CHICHESTER • BRISBANE • TORONTO • SINGAPORE

Library of Congress Cataloging in Publication Data:

Ludwig, John A.
    Statistical ecology: a primer on methods and computing / John A.
Ludwig, James F. Reynolds.
        p.   cm.
    "A Wiley-Interscience publication."
    Bibliography: p.
    Includes index.
    ISBN 0-471-83235-9
    1. Ecology—Statistical methods.   I. Reynolds, James F., 1946–
II. Title.
QH541.15.S72L83   1988
574.5′24′015195—dc 19   87-26348
                                                                        CIP

Printed in the United States of America

10 9 8 7 6 5 4 3 2

LIMITED USE LICENSE AGREEMENT

THIS IS THE JOHN WILEY & SONS, INC. (WILEY) LIMITED USE LICENSE AGREEMENT, WHICH GOVERNS YOUR USE OF THE WILEY PROPRIE-TARY SOFTWARE PRODUCTS (LICENSED PROGRAM) AND USER MAN-UAL(S), CONTAINED WITHIN IT.

YOUR OPENING OF THE SEALED DISK PACKAGE WHICH CONTAINS THE LICENSED PROGRAM INDICATES YOUR ACCEPTANCE OF THE TERMS AND CONDITIONS OF THIS AGREEMENT. IF YOU DO NOT AC-CEPT OR AGREE WITH THEM, YOU MUST RETURN THE LICENSED PRO-GRAM UNOPENED WITHIN 30 DAYS OF PURCHASE, AS EVIDENCED BY A COPY OF YOUR RECEIPT, AND THE PURCHASE PRICE WILL BE FULLY REFUNDED.

License: Wiley hereby grants you, and you accept, a non-exclusive and non-transferrable license, to use the Licensed Program and user manual(s) on the following terms and conditions:

a. The Licensed Program and User Manual(s) are for your personal use only.

b. You may use the Licensed Program on a single computer, or on its temporary replacement, or on a subsequent computer only.

c. You may modify the Licensed Program for your use only, but any such modifica-tions void all warranties expressed or implied. Otherwise, the modified programs will continue to be subject to the terms and conditions of this Agreement.

d. A backup copy or copies may be made only as provided by the user manual(s), but all such backup copies are subject to the terms and conditions of this Agreement.

e. You may not use the Licensed Program on more than one computer system, make or distribute unauthorized copies of the Licensed Program or User Manual(s), create by decompilation or otherwise the source code of the Licensed Program OR USE, COPY, MODIFY, OR TRANSFER THE LICENSED PROGRAM, IN WHOLE OR IN PART, OR USER MANUAL(S), EXCEPT AS EXPRESSLY PERMITTED BY THIS AGREEMENT.

IF YOU TRANSFER POSSESSION OF ANY COPY OR MODIFICATION OF THE LICENSED PROGRAM TO ANY THIRD PARTY, YOUR LI-CENSE IS AUTOMATICALLY TERMINATED. SUCH TERMINATION SHALL BE IN ADDITION TO AND NOT IN LIEU OF ANY EQUITABLE, CIVIL, OR OTHER REMEDIES AVAILABLE TO WILEY.

Term: This License Agreement is effective until terminated. You may terminate it any time by destroying the Licensed Program and User Manual together with all copies made (with or without authorization).

This Agreement will also terminate upon the conditions discussed elsewhere in this Agreement, or if you fail to comply with any term or condition of this Agreement. Upon such termination, you agree to destroy the Licensed Program, User Manual(s), and any copies made (with or without authorization) of either.

v

Wiley's Rights: You acknowledge that the Licensed Program and User Manual(s) are the sole and exclusive property of Wiley. By accepting this Agreement, you do not become the owner of the Licensed Program or User Manual(s), but you do have the right to use them in accordance with the provisions of this Agreement. You agree to protect the Licensed Program and User Manual(s) from unauthorized use, reproduction or distribution.

Warranty: To the original licensee only, Wiley warrants that the diskettes on which the Licensed Program is furnished are free from defects in materials and workmanship under normal use for a period of ninety (90 days from the date of purchase as evidenced by a copy of your receipt. If during the ninety day period, a defect in any diskette occurs, you may return it with proof of purchase. Wiley will replace the defective diskette(s) without charge to you. Your sole and exclusive remedy in the event of a defect is expressly limited to replacement of the defective diskette(s) at no additional charge. This warranty does not apply to damage or defects due to improper use or negligence.

THIS LIMITED WARRANTY IS IN LIEU OF ALL OTHER WARRANTIES, EXPRESSED OR IMPLIED, INCLUDING, WITHOUT LIMITATION, ANY WARRANTIES OF MERCHANTIBILITY OR FITNESS FOR A PARTICULAR PURPOSE.

EXCEPT AS SPECIFIED ABOVE, THE LICENSED PROGRAM AND USER MANUAL(S), ARE FURNISHED BY WILEY ON AN "AS IS" BASIS AND WITHOUT WARRANTY AS TO THE PERFORMANCE OR RESULTS YOU MAY OBTAIN BY USING THE LICENSED PROGRAM AND USER MANUAL(S). THE ENTIRE RISK AS TO THE RESULTS OR PERFORMANCE, AND THE COST OF ALL NECESSARY SERVICING, REPAIR, OR CORRECTION OF THE LICENSED PROGRAM AND USER MANUAL(S) IS ASSUMED BY YOU.

IN NO EVENT WILL WILEY BE LIABLE TO YOU FOR ANY DAMAGES, INCLUDING LOST PROFITS, LOST SAVINGS, OR OTHER INCIDENTAL OR CONSEQUENTIAL DAMAGES ARISING OUT OF THE USE OR INABILITY TO USE THE LICENSED PROGRAM OR USER MANUAL(S), EVEN IF WILEY OR AN AUTHORIZED WILEY DEALER HAS BEEN ADVISED OF THE POSSIBILITY OF SUCH DAMAGES.

General: If any of the provisions of this Agreement are invalid under any applicable statute or rule of law, they are to that extent deemed omitted.

This Agreement represents the entire Agreement between us and supercedes any proposals or prior agreements, oral or written, and any other communication between us relating to the subject matter of this Agreement.

This Agreement will be governed and construed as if wholly entered into and performed within the State of New York.

YOU ACKNOWLEDGE THAT YOU HAVE READ THIS AGREEMENT, AND AGREE TO BE BOUND BY ITS TERMS AND CONDITIONS.

*In memory of our fathers, Wilton and Henry,
and to our mothers, Esther and Anna*

Preface

Our goal in this primer is to provide the beginning student with an introduction to some of the current statistical topics in community ecology. The scope and depth of coverage we give to the various methodologies presented are the product of our years of experience in teaching students having little or no prior exposure to statistical ecology and having only a basic background in statistics. In a one-semester course for beginning students we do not attempt an exhaustive survey of the many methodologies available. Rather, our intention is to introduce students to a select range of topics, some of which are (1) historically important in ecology (e.g., polar ordination), (2) popular among ecologists (e.g., diversity indices), or (3) powerful statistical tools for analyzing ecological patterns (e.g., multidimensional scaling). It follows that some of the methods presented might fall into one of these groups, but not necessarily all three. For example, we present diversity indices mainly because of their widespread popularity; students will have to contend with the frequent appearance of diversity indices in the ecological literature, and, therefore, we attempt to cover some of their uses and misuses. The depth of treatment is intended to be a reasonable balance between oversimplification and excessive mathematical treatment.

Historically, ecology has evolved from being largely a descriptive discipline to its present state as a highly quantitative field. Consequently, researchers have had to learn new, and often complex, quantitative techniques. Although several recent books on quantitative ecology (e.g., Legendre and Legendre 1983, Orloci and Kenkel 1985, Pielou 1984) provide details on the theory and uses of specific statistical methods, we feel that there remains a need for a primer-level approach that has a fairly broad coverage of topics. This is where we have aimed this book.

Statistical Ecology: A Primer is organized into seven parts. In the first, we give an overview of our philosophy on data collection in community ecology and briefly review some aspects of sampling and the organization of data into a matrix. Sampling is, of course, a vital part of statistical ecology and there are many adequate sources available on this subject; our objective in Part I is to reemphasize its importance.

The remaining parts of the book cover the topics of spatial pattern analysis, species abundance relations, species overlap models, community classification and ordination, and, finally, what we call community interpretation. Each part begins with a chapter containing background information to provide the student with a broad perspective on the subject. Toward this goal, we include a table of selected references so that interested students can study *specific* examples of the use of certain methods in ecological research. We have students in our courses read and critique many such papers and believe that exposure to the literature is an invaluable part of the learning process.

The remaining chapters in each part cover specific methodologies and are organized as follows. First, we give some necessary preparatory information, including references to complementary sources. Next, the computational procedures involved are explained in a step-by-step outline fashion. We use this outline approach because the sample calculations that follow are given in the same order, allowing quick reference to the appropriate computational steps and equations. The sample calculations are very simple (using contrived data sets), since our aim is to help the student understand the computations involved. In some cases, this compromises simplicity with ecological–statistical reality; we make an effort throughout the text to point out where such compromises exist. Following these worked examples we provide further examples, using computer programs for the computations.

At the end of each methods chapter we provide a section on additional topics and a summary of our recommendations. The additional topics consist of brief descriptions and selected references intended to highlight various extensions of methods that have not been included in the earlier presentations. Students will usually not find these treatments to be self-contained; rather, our goal is to note certain problems or advances that may be of interest to students after they obtain a basic familiarity with each method. The references provide a guide to further study. Lastly, the summary and recommendations section is intended to highlight important conclusions about the use (and abuse) of specific techniques.

An important factor that has contributed to the increased use of quantitative methods in ecology is the wide accessibility of computers. We feel that computers can serve as valuable pedagogical tools in enhancing the learning experience for students. Hence, we provide numerous BASIC computer programs for use in conjunction with the examples in the text, as well as with the students' own data sets. We encourage students, however, to work through the simplified examples provided by hand *before* using these programs in order to understand the computations involved. These computer programs were written specifically for BASIC interpreters and compilers on microcomputers operating under MSDOS/PCDOS. Microcomputers are ideal for small data sets, like those we provide in this text, because they require little computation time. For large data sets there are numerous mathematical and statistical programs available on mainframe computers for many of the methods we cover.

We express our sincere gratitude to the many students at New Mexico State University, North Carolina State University, and San Diego State University who stimulated our thinking and motivated us to write this text. We wish to acknowledge numerous colleagues who helped us immensely by reading various drafts of chapters and providing us with critical comments, including Mike Austin, Peter Diggle, Harvey Gold, Stuart Hurlbert, Dennis Knight, Robert Knox, Bob Peet, Fred Smeins, and Brian Walker. Finally, our families deserve special thanks for their support, encouragement, and tolerance.

JOHN A. LUDWIG
JAMES F. REYNOLDS

Deniliquin, NSW, Australia
San Diego, California
February 1988

Contents

STATISTICAL ECOLOGY

PART ONE
ECOLOGICAL COMMUNITY
DATA

CHAPTER 1

Background

Statistical ecology encompasses numerous quantitative methodologies that deal with the exploration of *patterns* in biotic communities. These patterns are of many different types, including the spatial dispersion of a species "within" a community, relationships between many species "within" a community, and relationships among many species "between" communities. Hence, our definition of statistical ecology falls within the broader arena of what is popularly known as mathematical or quantitative ecology, which encompasses both population dynamics and community patterns.

1.1 EXPERIMENTAL VERSUS OBSERVATIONAL DATA

Ecological data in community ecology may be viewed as a product of either an *experimental* or *observational* approach (Goodall 1970). It is helpful to distinguish between these approaches in order to clarify differences between various types of ecological research and, hence, the types and limits of the data collected.

An experimental approach presupposes that the community is subject to experimental manipulation. That is, we can divide the community into replicate portions on which various treatments and controls can be imposed. Therefore, any differences detected in measured responses can be attributed to the experimental treatments. On the other hand, using an observational approach, we make measurements on the community over a range of conditions imposed by nature rather than by the researcher. This leaves us with two alternatives: (1) to study different samples obtained at the same time but under different conditions (e.g., phytoplankton sampling of inshore and off-shore waters of a lake) or (2) to study samples at the same place but at different times (e.g., samples of offshore phytoplankton taken during the summer and winter).

Community ecologists are often interested in obtaining information pertaining to a large number of variables in a community, but without imposing any manipulations on these variables. That is, we usually follow an observational approach, which is inductive, nonexperimental, and multivariate (Noy-Meir 1970). As will be evident from our choice of topics and examples throughout the text, this is the approach we assume in this primer.

Much work in community ecology is motivated by a desire to elucidate and describe patterns in our data sets, rather than formally testing a priori hypotheses (Green 1980). This is an important distinction that needs to be emphasized often. We consider most of what we cover in this primer to fall under the heading of *pattern detection methods*. As Greig-Smith (1971) notes, observations in community ecology are usually made over space and/or time with a variety of possible goals, including, for example, estimating the overall species composition within an area, correlating species properties with environmental factors, and studying temporal and spatial variability in species patterns. In some cases, the detection of specific patterns across samples may lead to the formation of causal hypotheses about the underlying structure of the ecological community (Noy-Meir 1970), which may then be tested with further work (perhaps using experimental approaches).

1.2   ECOLOGICAL SAMPLING

The various stages of an observational study are shown in Figure 1.1. The first step involves a clear definition of the aims of the study. We want to emphasize the importance of this step, since it influences all subsequent aspects of data collection and analyses. This first step defines the scope or domain of the study.

The selection of an appropriate sampling scheme is also important. As Greig-Smith (1983) states: "the value of quantitative data ... depends on the sampling procedure used to obtain them." Of course, any scheme adopted must be related to the aims of the study and should be designed to maximize the information obtained in return for the effort and time invested. Although we do not present specific guidelines on sampling *per se*, we attempt throughout the text to highlight instances where sampling problems might occur and to stress the importance of sampling in certain analyses. For excellent introductions to sampling methods and theories appropriate for community ecology, we refer the student to one (or more) of the references cited in Section 1.5.

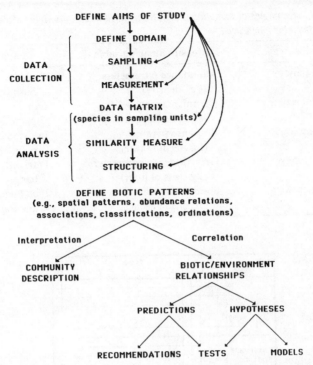

Figure 1.1   *Stages of an observational ecology approach (after Noy-Meir 1970).*

A successful sampling scheme involves the selection of an appropriate *sampling unit* (SU). Common sampling units used in ecology include quadrats, leaves of a plant, light traps, soil cores, pit traps, individual organisms, and belt transects (Table 1.1). Some SUs occur naturally (e.g., plant leaves), while others are arbitrarily defined (e.g., quadrats)—an important distinction discussed in Section 2.1. We will be referring to various types of SUs in the examples presented in the text. Note that it is incorrect to call the sampling units *samples*; a sample consists of a *collection* of sampling units (Pielou 1974). For example, using a 1 m² SU, a sample might consist of 20 such SUs randomly located throughout the study area.

Once the sampling procedure and the choice of SU have been made, specific measurements (e.g., presence–absence, biomass, density) are taken on the species of interest in the community. These data are then tabulated into an *ecological data matrix*, which is a convenient method of summarizing large data sets and is the basic unit that we subject to analyses. The data matrix is a rectangular display of the measurements taken in each SU. There are two

TABLE 1.1   *Examples of sampling units (SUs) used in various fields of ecology and typical variables measured*

Field	Measurement Variable	Sampling Unit
Population ecology	Abundance parameters (e.g., density)	Trapline, grid, or flush transect
Physiological ecology	Organism response (e.g., photosynthesis)	Individual animal, plant, or microbe
Behavioral ecology	Response parameters (e.g., feeding rate)	Individual animal, plant, or microbe
Microbial ecology	Growth/decay parameters (e.g., loss of litter)	Petri dish or litter-bag
Community ecology	Species abundances or presence/absence	Plot, quadrat, or line transect

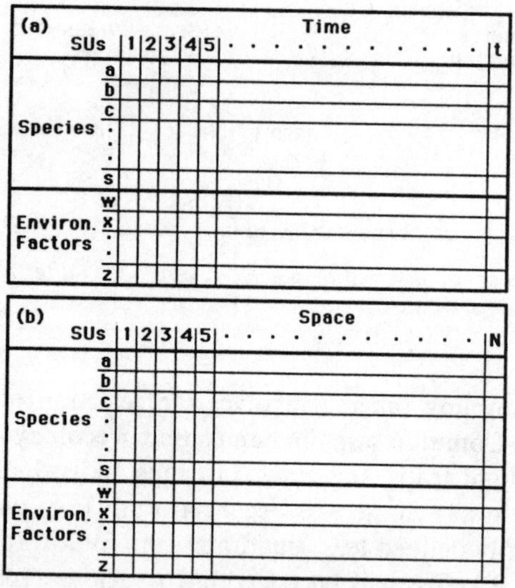

Figure 1.2   *Two basic types of ecological data matrices: (a) species and environmental factor data measured in one location through time (a total of t observations, the SUs) and (b) species and environmental factor data measured on N SUs over space, that is, at different locations in the landscape.*

basic types of data matrices depending on the purpose of study. First, in studies dealing with *temporal dynamics* (e.g., community succession), the data matrix represents measurements made on species through *time* (Figure 1.2a). The time interval depends on the specific purpose of the study. Environmental factors (e.g., soil water content, soil pH, soil temperature) are often simultaneously collected in the SUs. The second type of data matrix deals with measurements taken on a number of SUs distributed over *space* (Figure 1.2*b*). The actual spatial distribution of the SUs is determined by the experimental design (e.g., random placement of quadrats). Of course, studies over space may be repeated seasonally or yearly, for example, to examine spatial dynamics through time. In the background chapter beginning each part of the text, we show the specific form of the ecological data matrix being analyzed.

1.3   ECOLOGICAL DATA AND MICROCOMPUTERS

The BASIC computer programs provided with this text were written for use on microcomputers. We designed these programs to aid the student in comprehending the various examples in the text and for experimenting with his or her own data. Compared to large mainframe computers, microcomputers have the obvious limitations of available memory and speed of computation and, consequently, these programs are not intended for use with the type of large data sets common to many community ecology studies. For the analysis of large data sets, mainframe computers are, of course, both efficient and time-economical. Software packages that will perform on large computers many of the statistical techniques covered in this primer are available for use by ecologists (see Section 1.5).

Numerous pitfalls potentially exist when using "canned" programs on any type of computer. The user can do many types of analyses with, literally, a press of a button. This requires no knowledge of the computations and assumptions involved, often leading to the inappropriate application of a method and then misinterpretation of the results. Also, there may be either computational or logical programming errors that go undetected with test data but, nevertheless, produce incorrect results with other data sets (such as yours!). For example, in an early version of one of our programs correct results were obtained as long as the ecological data matrix consisted of more rows (species) than columns (SUs). It just happened that all test data we used as standards in developing this particular program happened to have more rows than columns. Of course, the first student who ran this program happened to use a data set with fewer rows than columns and erroneous results were obtained.

Round-off errors are also a problem which may arise in attempting to interpret output from microcomputer programs. Computers store and compute with a certain fixed number of digits, the maximum value of which varies with the size of the computer. Consequently, using the same program on different computers with various capacities may produce different results attributable to such round-off errors. The student should be aware of such possibilities. In-depth discussions of round-off error in computers can be found in Nelder (1975) and Lewis and Doerr (1976).

1.4 STATISTICAL DEFINITIONS

Some statistical terms that we frequently use throughout the text are defined below.

1. *Population*—the entire collection of individual SUs that are potentially observable in an ecological community. It is from this population that a sample will be drawn and statistical inferences made.
2. *Parameter*—a known characteristic of a population (e.g., density).
3. *Statistic*—an estimator of a population parameter (e.g., the expected population density obtained by sampling the population).
4. *Accuracy*—a measure of the closeness of a statistic obtained using a certain sampling procedure to the true value of a population parameter.
5. *Precision*—a measure of the degree of repeatability of a statistic in replicated samples using a certain sampling procedure. The standard error is considered the basic expression of sampling precision. Note that it is possible to have a precise estimate without accuracy, since a particular sampling procedure may be precise but biased.
6. *Bias*—a consistent under- or overestimate of the true population parameter by a statistic. Statistical bias is usually a result of a consistent inaccuracy in a sampling procedure or, in some cases, a formula.
7. *Error*—the difference between a statistic and the true value of a population parameter. It is important to realize that bias and error may arise both in the original collection of data and in the subsequent manipulation of the data.

1.5 SUMMARY AND RECOMMENDATIONS

1. Much work in community ecology is motivated by the desire to elucidate and describe *patterns* rather than formally testing hypotheses. This is an important distinction to keep in mind throughout this primer (Section 1.1).

2. The first step in any study is to have a clear definition of its goals (Section 1.2).

3. An important step in designing a field study (or interpreting published ones) is to identify the appropriate *sampling unit* (Section 1.2 and Table 1.1). A *sample* is obtained using these SUs.

4. Some excellent references for overviews on sampling considerations in ecological studies include Chapter 2 in Greig-Smith (1983), Chapter 2 in Southwood (1978), Chapters 2 and 3 in Green (1979), Cochran (1963), Williams (1976), Eberhardt (1976), Chapters 1 and 4 in Cox (1985), Appendices 1–9 in Myers and Shelton (1980), Chapter 4 in Mueller-Dombois (1974), and Knapp (1984).

5. Computers are useful tools for analyzing data; however, be cognizant of the potential pitfalls in using "canned" programs (Section 1.3). Numerous software packages for use on large mainframe computers are available for performing most of the analyses presented in this primer. These include BMDP (Dixon and Brown 1979), ORDIFLEX (Gauch 1977), SAS (Ray 1982), NT/SYS (Rohlf et al. 1971), CLUSTAR/CLUSTID (Romesburg 1984), CLUSTAN (Wishart 1969), and programs in books by Cooley and Lohnes (1971), Orloci (1978, 1985), and Williams (1976). Green (1979) and Legendre and Legendre (1983) review some of these packages.

PART TWO
SPATIAL PATTERN ANALYSIS

CHAPTER 2

Background

The spatial pattern of plants and animals is an important characteristic of ecological communities. This is usually one of the first observations we make in viewing any community and is one of the most fundamental properties of any group of living organisms (Connell 1963).

Three basic types of patterns are recognized in communities: random, clumped, and uniform (Figure 2.1). In this part of the book we present various methods of detecting spatial patterns. Once a pattern has been identified, the ecologist may propose and test hypotheses that explain what underlying causal factors may be responsible. Hence, the ultimate objective of detecting spatial patterns is to generate hypotheses concerning the structure of ecological communities (Williams 1976). For example, this was the case in a study by George and Edwards (1976), who investigated the importance of wind-induced water movement on the nonrandom, horizontal patterning of phytoplankton and zooplankton in a small reservoir in Wales. In another study, also in Wales, Doncaster (1981) found that the clumped pattern of ant nests on a large island was largely determined by the influence of slope exposure, moisture, and rabbit grazing on the island vegetation. As a final example, LaMont and Fox (1981) studied the spatial patterns of some acacia trees in Western Australia at two levels of intensity, between and within clumps of trees; they found that the pattern at both levels was affected by drought and by selective animal grazing.

The following causal mechanisms are often used to explain observed patterns in ecological communities (Pemberton and Frey 1984). Random patterns in a population of organisms imply environmental homogeneity and/or nonselective behavioral patterns. On the other hand, nonrandom patterns

13

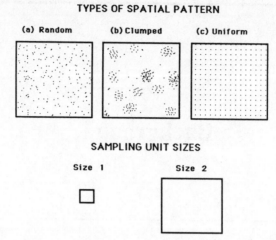

Figure 2.1 *Three types of spatial pattern: (a) random, where all individuals are located independently of each other; (b) clumped, where individuals tend to be located together in clusters; and (c) uniform, where individuals are regularly spaced. Two SU sizes are shown.*

(clumped and uniform) imply that some constraints on the population exist. Clumping suggests that individuals are aggregated in more favorable parts of the habitat; this may be due to gregarious behavior, environmental heterogeneity, reproductive mode, and so on. Uniform dispersions result from negative interactions between individuals, such as competition for food or space. Of course, detecting a pattern and explaining its possible causes are separate problems. Furthermore, it should be kept in mind that nature is multifactorial; many interacting processes (biotic and abiotic) may contribute to the existence of patterns (Quinn and Dunham 1983).

Hutchinson (1953) was one of the first ecologists to consider the importance of spatial patterns in communities and identified various causal factors that may lead to patterning of organisms: (1) vectorial factors resulting from the action of external environmental forces (e.g., wind, water currents, and light intensity); (2) reproductive factors attributable to the reproductive mode of the organism (e.g., cloning and progeny regeneration); (3) social factors due to innate behaviors (e.g., territorial behavior); (4) coactive factors resulting from intraspecific interactions (e.g., competition); and (5) stochastic factors resulting from random variation in any of the preceding factors. Thus, processes contributing to spatial patterns may be considered as either intrinsic to the species (e.g., reproductive, social, and coactive) or extrinsic (e.g., vectorial). Further discussions of the causes of pattern are given in Kershaw (1973), Southwood (1978), and Greig-Smith (1979, 1983).

Plant community ecologists have largely been responsible for the development of many of the methods of spatial pattern analysis (SPA). This can be attributed mainly to the sessile nature of plants and the more obvious nature of patterns in plant communities. SPA is usually restricted to small-scale patterns within a community. Of course, the choice of scale is important. Various degrees of species clumping occur at intervening scales in the community; pattern may be detected at one scale but not at another. In fact, some SPA techniques of pattern detection are based on exploring for pattern at various scales in the community. Pielou (1979) gives an example of the application of SPA to large-scale biogeographical studies.

The SPA methods we present deal exclusively with the detection of patterns in space. However, it is common to repeat studies of pattern analysis over different seasons or years to explore patterns through time.

2.1 MATRIX VIEW

SPA models are based on data from measurements of abundance of a species across sampling units (SUs) (Chapters 3 and 4) or distances between individuals within the community (Chapter 5). The matrix view of species across SUs is shown in Figure 2.2. The nature of the SU has very important implications when interpreting the results of SPA studies; these implications cause some problems (described in Chapters 3 and 4). Because of these problems, SPA techniques using distances have certain advantages, which we cover in Chapter 5.

Pielou (1977, 1979) distinguishes between *natural* and *arbitrary* SUs. We can define natural SUs for organisms that occur in discrete segments of a habitat; for example, termites in decaying logs, insects found on fruits, or detritivores in dung pats. The logs, fruits, and pats are considered to be natural SUs. Various statistics and indices based on the abundance data of organisms

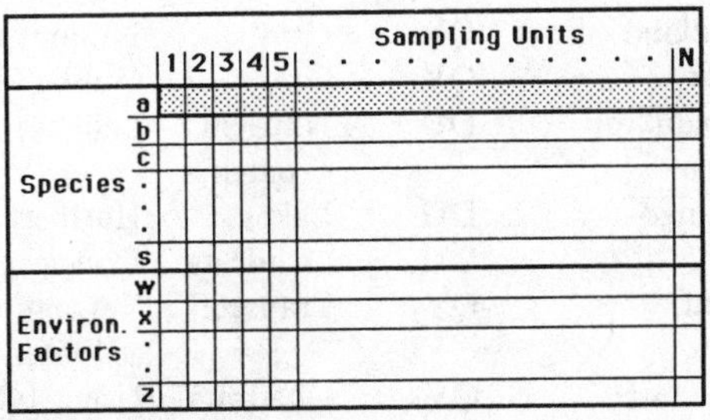

Figure 2.2 The shaded area indicates the form of the ecological data matrix for SPA. Interest is in the dispersion of individuals of a single species (e.g., species a) across many (N) SUs. The SUs are either natural (e.g., pine cones) or arbitrary (e.g., quadrats).

obtained from such natural SUs may convey meaningful information about spatial pattern (Pielou 1979).

For organisms that occur in continuous habitats, such as trees in a forest, zooplankton in a lake, and grasses in a prairie, it is necessary to use some arbitrary SU (e.g., plots or quadrats) to obtain a sample. Consequently, a sample statistic like the mean number of trees per plot is less helpful in conveying meaningful information about spatial pattern because different sizes of the SU may result in different conclusions. This is depicted in Figure 2.1, where it can be seen that repeated random sampling of the clumped population with SU of size 1 will yield many SUs with zero or few individuals and, occasionally, many individuals. This would suggest to us that the population is clumped. However, using an SU of size 2 will tend to give about the same number of individuals per SU for all SUs, suggesting a uniform, rather than clumped, pattern! It can be seen from this example that the detection of pattern in continuous habitats will depend on the scale of the pattern relative to the size of the SU; the problem is that we are unlikely to know what this relationship is before sampling. For excellent discussions on the relationships among sampling, sampling units, and spatial pattern we recommend Pielou (1977).

TABLE 2.1   *Selected literature for SPA models*

Location	Community	Method[a]	SU[b]	Reference
Michigan	Forest	DS	Points	Dice 1952
Michigan	Forest	QV	Grid	Squires and Klosterman
		DS	Points	1981
England	Grassland	DU	Quadrats	Clapham 1936
Australia	Salt marsh	QV	Quadrats	Goodall 1963
California	Desert	DS	Points	Gulmon and Mooney 1977
Wales	Lake	DU	Bottles	George and Edwards 1976
Canada	Grass–legume pasture	QV	Quadrats	Turkington and Cavers 1979
Australia	Woodland	QV	Grid	Lamont and Fox 1981
Costa Rica	Forest	QV	Grid	Hubbell 1979
Laboratory	Zooplankton algae	DU	Nutrient patches	Lehman and Scavia 1982
Andes	Flamingo	DU	Lakes	Hurlbert and Keith 1979
Wales	Ant	DU	Quadrats	Doncaster 1981
Canada	Lichen	QV	Transect	Yarranton and Green 1966
Subantarctic	Moss–turf	QV	Quadrats	Usher 1983

[a]DU = distribution models; QV = quadrat variance models; and DS = distance models.
[b]Note the type of sampling unit (SU) used in each study.

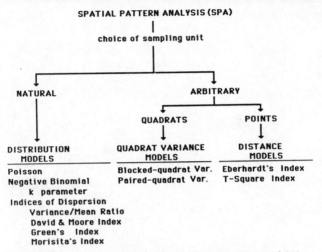

Figure 2.3   Types of SPA models following choice of SU.

2.2   SELECTED LITERATURE

Some select examples of ecological applications of SPA are given in Table 2.1. Depending on the choice of SU, pattern detection studies can be placed into one of three distinct categories of SPA models. These are illustrated in Figure 2.3. *Distribution models*, presented in Chapter 3, are based on the observed relative frequency distributions of species abundance data across the natural SUs of the sample. Also based on relative frequency distributions are a number of indices of dispersion that measure the degree of clumping. *Quadrat variance models*, covered in Chapter 4, are applicable when observations come from arbitrary SUs. *Distance models*, illustrated in Chapter 5, apply when observations are from distances between points and individuals. Many ecologists have routinely applied distribution models to data obtained with the use of arbitrary SUs in spite of the known dependency of SPA results and conclusions on the size and shape of the SU, as discussed in Section 2.1. The reason this is incorrect will be clearer after the student reads Chapter 3, but we note in passing that these are the types of errors we want to avoid.

CHAPTER 3

Distribution Methods

In this chapter we describe the use of *statistical distributions* and *indices of dispersion* for detecting and measuring spatial pattern of species in communities. If individuals of a species are spatially dispersed over discrete sampling units (SUs), e.g., scale insects on plant leaves, and if at some point in time a sample is taken of the number of individuals per SU, then it is possible to summarize these data in terms of a frequency distribution. This frequency distribution consists of the number of SUs with 0 individuals, 1 individual, 2 individuals, and so on. This constitutes the basic data set we use in the *pattern detection methods* described subsequently. Note that the species are assumed to occur in discrete sites or natural SUs such as leaves, fruits, trees (see Section 2.1).

3.1 GENERAL APPROACH

What will the observed frequency distribution of the number of individuals per SU tell us about a spatial pattern? As an example, consider some possible ways that scale insects could be dispersed among leaves of a plant. As illustrated in Figure 3.1, a total of 30 insects were apportioned in a *random*, *clumped*, and *uniform* manner among 10 leaves on each of three plants. Of course, the mean number of individuals per leaf (SU) is 3 for each plant, regardless of the way they are dispersed.

In the case of random dispersal, each leaf has an equal chance of hosting an individual, and the presence of an individual on a leaf does not influence whether or not other individuals will be present. The frequency distribution

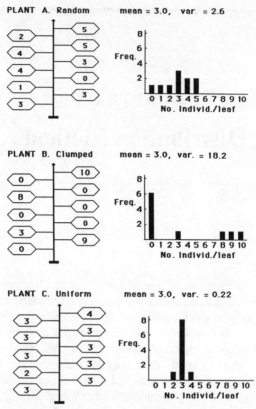

Figure 3.1 *Possible dispersals for 30 scale insects on 10 leaves (natural SUs) showing number of insects found on each leaf. The mean and variance for the number of insects per leaf for plants A (random), B (clumped), and C (uniform) is given.*

for the random pattern shows a peak at 3 individuals per leaf (the mean) with a spread on either side of the peak. In the clumped pattern there are numerous leaves with no individuals and a few with many individuals. This also can be contrasted to the uniform pattern where most of the leaves have 3 individuals.

Also, note how the variance of the number of individuals per leaf is strongly affected by the pattern of dispersal of the insects. In the random pattern (plant *A*), the resultant variance (2.6) is close to the mean. Where individuals are clumped or aggregated on a few leaves (plant *B*), the variance (18.2) is much greater than the mean. In the case where the number of individuals per leaf is nearly uniform (plant *C*), the resultant variance (0.22) is much less than the mean. (Of course, had each leaf contained exactly the same number of individuals, the variance would be zero.)

In summary, the relationships between the mean and variance of the number of individuals per SU is influenced by the underlying pattern of dispersal of the population. We can now define the three basic types of patterns and their variance-to-mean relationships, where σ^2 = variance and μ represents the mean:

1. Random pattern: $\sigma^2 = \mu$
2. Clumped pattern: $\sigma^2 > \mu$
3. Uniform pattern: $\sigma^2 < \mu$.

There are certain statistical frequency distributions that, because of their variance-to-mean properties, have been used as models of these types of ecological patterns:

1. *The Poisson distribution* ($\sigma^2 = \mu$) for random patterns
2. *The negative binomial* ($\sigma^2 > \mu$) for clumped patterns
3. *The positive binomial* ($\sigma^2 < \mu$) for uniform patterns.

While these three statistical models have commonly been used in studies of spatial pattern, it should be recognized that other statistical distributions might be equally appropriate (Pielou 1977).

The initial step in pattern detection in community ecology often involves testing the hypothesis that the distribution of the number of individuals per SU is random. If this hypothesis is rejected, then the distribution may be in the direction of clumped (usually) or uniform (rarely). If the direction is toward a clumped dispersion, agreement with the negative binomial may be tested and certain indices of dispersion, which are based on the ratio of the variance to mean, may be used to measure the degree of clumping. Because of the relative rarity of uniform patterns in ecological communities, this case is not considered here. For those interested, Elliott (1973) gives some ecological examples of uniform patterns.

Before proceeding, we wish to make some cautionary points. First of all, failure to reject a hypothesis of randomness means only that we have failed to detect nonrandomness using the specific data set at hand. Would an independent data set collected in the same community lead to a similar conclusion? Second, we should propose only reasonable hypotheses (Pielou 1974): a hypothesis should be tenable and based on a mixture of common sense and biological knowledge. This second point has important ramifications with regard to the first. It is not uncommon for a theoretical statistical distribution (e.g., the Poisson series) to resemble an observed frequency distribution (i.e., there is a statistical agreement between the two) even though the assumptions underlying this theoretical model are not satisfied by the data set. Consequently, we may accept a null hypothesis that has, in fact, no biological justification. Third, we should not base our conclusions on significance tests alone. All available sources of information (ecological and statistical) should be used in concert. For example, failure to reject a null hypothesis that is based on a small sample size should be considered only as a weak confirmation of the null hypothesis (Snedecor and Cochran 1973). Lastly, remember that the detection of spatial pattern (our objective here) and explaining its possible causal factors are separate problems (see Chapter 2).

3.2 PROCEDURES

3.2.1 Poisson Probability Distribution

For a randomly dispersed population of organisms, the Poisson model ($\sigma^2 = \mu$) gives probabilities for the number of individuals per SU, provided the following conditions hold: (1) each natural SU has an equal probability of hosting an individual, (2) the occurrence of an individual in a SU does not influence its occupancy by another, (3) each SU is equally available, and (4) the number of individuals per SU is low relative to the maximum possible that could occur in the SU (see Greig-Smith 1983, pp. 58–59). To compute these probabilities, we need only an estimate of the mean number of individuals per SU (μ). The steps in this procedure are outlined below:

STEP 1. STATE THE HYPOTHESIS. The hypothesis is that the number of individuals per SU are from a Poisson distribution. If this hypothesis is not rejected, we may conclude that either the population is randomly dispersed or a nonrandom pattern does, in fact, exist but this test did not detect it (see Section 3.1). If the hypothesis is rejected, the direction of the nonrandom pattern may be of interest and other SPA models may be used (some possibilities are discussed later in the chapter).

STEP 2. THE FREQUENCY DISTRIBUTION, F_x. The sample data, which consist of the number of individuals per SU, are summarized as a frequency distribution, that is, the number of SUs with $x = 0, 1, 2, \ldots, r$ individuals. It should be noted here that a fairly large sample size is necessary to arrange data in this fashion. A rule of thumb is that the minimum number of SUs (N) should be greater than 30; if N is less than this, see Section 3.2.3.

STEP 3. THE POISSON PROBABILITIES, $P(x)$. The probability of finding x individuals in a SU, that is, $P(x)$, where $x = 0, 1, 2, \ldots, r$ individuals, is given by the Poisson series

$$P(x) = (\mu^x e^{-\mu})/x! \tag{3.1}$$

where e is the base of the natural logarithms (2.7183) and $x!$ is the factorial of x [for example, for $x = 3$, $x! = (3)(2)(1) = 6$]. The mean (μ) is the only parameter in the Poisson model. To estimate μ, the statistic $\bar{x}$ is computed from the data as the mean number of individuals per SU. Using $\bar{x}$ as an estimate of μ, the probabilities of $x = 0, 1, 2, \ldots, r$ individuals per SU are

$$P(0) = e^{-\bar{x}}$$

$$P(1) = (\bar{x})^1 e^{-\bar{x}}/1! \quad \text{or} \quad (\bar{x}/1)P(0)$$

$$P(2) = (\bar{x})^2 e^{-\bar{x}}/2! \quad \text{or} \quad (\bar{x}/2)P(1)$$

$$\vdots$$

$$P(r) = (\bar{x})^r e^{-\bar{x}}/r! \quad \text{or} \quad (\bar{x}/r)P(r-1)$$

Since these are probabilities, the sum of the $P(x)$'s will be one. Note that once $P(0)$ is computed, the simplified form of computing these probabilities (shown on the far right-hand side of the preceding equations) eliminates the need to compute factorials.

STEP 4. THE EXPECTED POISSON FREQUENCIES, E_x. The Poisson model is a probability distribution, and when each probability is multiplied by the total number (N) of SUs in the sample, the expected number of SUs containing 0, 1, 2, \ldots, r number of individuals can be determined. Thus, letting E_x represent the expected frequencies of $x = 0, 1, 2, \ldots, r$ individuals per SU, we have

$$E_0 = (N)P(0)$$

$$E_1 = (N)P(1)$$

$$E_2 = (N)P(2)$$

$$\vdots$$

$$E_r = (N)P(r)$$

Each of these equations represents a frequency class, resulting in a total of $q = r + 1$ frequency classes of expected individuals. If the probabilities in the tails of the distribution drop so low that the expected number of individuals becomes quite small, these frequency classes are pooled using the following rules (Sokal and Rohlf 1981): (1) if q is less than 5, then the lowest expected value should not be less than 5 individuals or (2) if q is greater than or equal to 5, then the lowest expected value should not be less than 3 individuals. As a general rule, no expected value (E_x) should ever be less than 1 (Snedecor and Cochran 1973) or, more rigorously, much less than 5 (Poole 1974). These are not strict rules and minimum expected values of 1 and 3 may be tried. In cases were pooling significantly affects the results of the test, we recommend the more conservative values. The number of frequency classes that remain after any necessary pooling is the new value for q.

STEP 5. GOODNESS-OF-FIT TEST STATISTIC, χ^2. The chi-square goodness-of-fit test is used to ascertain how well the observed frequencies (F_x, step 2) compare to the expected frequencies (E_x, step 4). The chi-square test statistic (χ^2) is computed as

$$\chi^2 = \sum_{x=0}^{q} [(F_x - E_x)^2 / E_x] \qquad (3.2)$$

This test statistic is compared to a table of chi-square probabilities with $q - 2$ degrees of freedom (recall that $q = r + 1$, step 4). If the chi-square test statistic is greater than the tabular chi-square at some selected level of probability (e.g., 5%), we conclude that it is improbable that the frequency distribution follows a Poisson series and, hence, reject the null hypothesis.

3.2.2 Negative Binomial Distribution

The negative binomial model is probably the most commonly used probability distribution for clumped, or what often are referred to as *contagious* or *aggregated* populations (Sokal and Rohlf 1981). When two of the conditions associated with the use of Poisson model are not satisfied (Section 3.2.1), that is, condition 1 (each natural SU has an equal probability of hosting an individual) and condition 2 (the occurrence of an individual in a SU does not influence its occupancy by another), it usually leads to a high variance-to-mean ratio of the number of individuals per SU. As previously shown, this suggests that a clumped pattern may exist.

The negative binomial has two parameters: (1) μ, the mean number of individuals per SU and (2) k, a parameter related to the degree of clumping. The steps in testing the agreement of an observed frequency distribution with the negative binomial are outlined below; steps similar to those described for the Poisson series (Section 3.2.1) are cross-referenced rather than repeated.

STEP 1. STATE THE HYPOTHESIS. The hypothesis to be tested is that the number of individuals per SU follows a negative binomial distribution, and, hence, a nonrandom or clumped pattern exists. Failing to reject this hypothesis, the ecologist may have a good empirical model to describe a set of observed frequency data; this does not explain what underlying causal factors might be responsible for the pattern. Bliss and Calhoun (1954) discuss a number of naturally occurring phenomena that could give rise to a negative binomial distribution during the dispersal of a population of organisms. For example, if a population of insects randomly deposited egg clusters on the leaves of plants (a Poisson process) and if the number of larvae that hatched per cluster were independently distributed as a logarithmic distribution, the frequency counts of the number of larvae per leaf would follow a negative binomial probability distribution. While such a scenario might be possible, Bliss and Fisher (1953), Pielou (1977), and Solomon (1979) give numerous examples of how *contradictory hypotheses* regarding possible causal mechanisms (such as the preceding scenario) can, in fact, lead to the *same* probability distribution. This should always be kept in mind when using distribution models. We should not attempt to infer causality solely based on our pattern detection approaches.

STEP 2. *THE FREQUENCY DISTRIBUTION, F_x.*   As in Section 3.2.1, the number of individuals per SU is summarized as a frequency distribution, that is, the number of SUs with 0, 1, 2, ..., r individuals.

STEP 3. *THE NEGATIVE BINOMIAL PROBABILITIES, $P(x)$.*   The probability of finding x individuals in a SU, that is, $P(x)$, where $x = 0, 1, 2, ..., r$ individuals, is given by

$$P(x) = [\mu/(\mu + k)]^x \{(k + x - 1)!/[x!(k - 1)]!\} [1 + (\mu/k)]^{-k} \qquad (3.3)$$

The parameter μ is estimated from the sample mean ($\bar{x}$). Parameter k is a measure of the degree of clumping and tends toward zero at maximum clumping (see Section 3.6). An estimate for k (i.e., $\hat{k}$) is obtained using the following iterative equation:

$$\log_{10}(N/N_0) = \hat{k} \log_{10}[1 + (\bar{x}/\hat{k})] \qquad (3.4)$$

where N is the total number of SUs in the sample and N_0 is the number of SUs with 0 individuals. First, an initial estimate of $\hat{k}$ is substituted into the right-hand side (RHS) of this equation and the resultant value is compared to the value of the left-hand side (LHS). If the RHS is lower than the LHS, a higher value of $\hat{k}$ is then tried, and, again, the two sides are compared. This process is continued in an iterative fashion (appropriately selecting higher or lower values of $\hat{k}$) until a value of $\hat{k}$ is found such that the RHS converges to the same value as the LHS. A good initial estimate of $\hat{k}$ for the first iteration is obtained from

$$\hat{k} = \frac{\bar{x}^2}{s^2 - \bar{x}} \qquad (3.5)$$

where s^2 is the sample estimate of variance.

When the mean is small (less than 4), Eq. (3.4) is an efficient way to estimate $\hat{k}$. On the other hand, if the mean is large (greater than 4), this iterative method is efficient only if there is extensive clumping in the population. Thus, if both the population mean ($\bar{x}$) and the value of $\hat{k}$ [the "clumping" parameter, as computed from Eq. (3.5)] are greater than 4, Eq. (3.5) is actually preferred over Eq. (3.4) for estimating $\hat{k}$ (Southwood 1978).

Once the two statistics, $\bar{x}$ and $\hat{k}$, are obtained, the probabilities of finding x individuals in a SU, that is, $P(x)$, where $x = 0, 1, 2, \ldots, r$ individuals, are computed from Eq. (3.3) as

$$P(0) = [\bar{x}/(\bar{x} + \hat{k})]^0[(\hat{k} + 0 - 1)!/(0!(\hat{k} - 1)!)][1 + (\bar{x}/\hat{k})]^{-k}$$
$$= [1 + (\bar{x}/\hat{k})]^{-k}$$

$$P(1) = [\bar{x}/(\bar{x} + \hat{k})]^1[(\hat{k} + 1 - 1)!/(1!(\hat{k} - 1)!)][1 + (\bar{x}/\hat{k})]^{-k}$$
$$= [\bar{x}/(\bar{x} + \hat{k})](\hat{k}/1)P(0)$$

$$P(2) = [\bar{x}/(\bar{x} + \hat{k})]^2[(\hat{k} + 2 - 1)!/(2!(\hat{k} - 1)!)][1 + (\bar{x}/\hat{k})]^{-k}$$
$$= [\bar{x}/(\bar{x} + \hat{k})][(\hat{k} + 1)/2]P(1)$$

$$\vdots$$

$$P(r) = [\bar{x}/(\bar{x} + \hat{k})]^r[(\hat{k} + r - 1)!/(r!(\hat{k} - 1)!)][1 + (\bar{x}/\hat{k})]^{-k}$$
$$= [\bar{x}/(\bar{x} + \hat{k})][(\hat{k} + r - 1)/r]P(r - 1)$$

STEP 4. *THE EXPECTED NEGATIVE BINOMIAL FREQUENCIES, E_x.*   As was done for the Poisson model in Section 3.2.1, the expected number of SUs containing x individuals is obtained by multiplying each of the negative binomial probabilities by N, the total number of SUs in the sample. The number of frequency classes, q, is also determined as described for the Poisson model.

STEP 5. *GOODNESS-OF-FIT TEST STATISTIC, χ^2.*   The chi-square test statistic (χ^2) is computed using Eq. (3.2) and compared to a table of chi-square probabilities with $q - 2$ (parameters) $- 1$, that is, $q - 3$ degrees of freedom.

3.2.3 Indices of Dispersion

As described in Section 3.2.1, the variance and mean are equal in a theoretical Poisson distribution. Because of this, a number of indices based on variance-to-mean ratios have been proposed to (1) test the *equality* of the variance to mean in a Poisson series and (2) measure the *degree of clumping* of a population of organisms. In this section we present three such indices: the *index of dispersion*, the *index of clumping*, and *Green's index*.

Before proceeding, we want to make two points. First, the variance-to-mean ratio is usually referred to as the *index of dispersion* in the ecological literature (as in this chapter). However, over the years, a large number of variants of this ratio (and others) have been proposed to measure the degree of clumping; these ratios are called collectively "indices of dispersion" (including the index of clumping and Green's index). Second, recall that we are still considering the question of spatial pattern based on the number of individuals per SU, where the SUs are discrete, naturally occurring entities (Section 2.1). In Chapter 5 we present another index of dispersion, which is based on distances between organisms.

Index 1:   Index of Dispersion. The variance-to-mean ratio or index of dispersion (ID) is

$$ID = \frac{s^2}{\bar{x}} \tag{3.6}$$

where $\bar{x}$ and s^2 are, as before, the sample mean and variance. If the sample is in agreement with a theoretical Poisson series, we would expect this ratio to be equal to 1.0. Therefore, we can test for significant departures of ID from 1.0 using the chi-square test statistic

$$\chi^2 = \left(\sum_{i=1}^{N} (x_i - \bar{x})^2 \right) \Big/ \bar{x} \tag{3.7a}$$

$$= ID(N - 1) \tag{3.7b}$$

where x_i is the number of individuals in the ith SU and N is the total number of SUs. For small sample sizes ($N < 30$), χ^2 is a good approximation to chi-square with $N - 1$ degrees of freedom [in view of this, note the similarity of Eq. (3.7a) with Eq. (3.2)]. If the value for χ^2 falls between the chi-square table values at the 0.975 and 0.025 probability levels ($P > 0.05$), agreement with a Poisson (random) distribution is accepted (i.e., $s^2 = \bar{x}$). On the other hand, values for χ^2 less than the 0.975 probability level suggest a regular pattern (i.e., $s^2 < \bar{x}$), whereas χ^2 values greater than the 0.025 probability level suggest a clumped pattern (i.e., $s^2 > \bar{x}$). In Table 3.1, the ID and the associated χ^2 test statistic for each of the scale-insect dispersions shown in Figure 3.1 (where $N = 10$) are given. In each case, the test confirms the underlying pattern. For example, the hypothesis that the value for ID of 0.87 in plant A is equal to 1.0 would be accepted (i.e., the value of $\chi^2 = 7.8$ is within the range of the chi-square table values of 2.7–19.0).

For large sample sizes ($N \geq 30$), $\sqrt{(2\chi^2)}$ tends to be normally distributed, and another test statistic, d, may be computed as

$$d = \sqrt{2\chi^2} - \sqrt{2(N - 1) - 1} \tag{3.8}$$

TABLE 3.1 *Values for the index of dispersion (ID) and Green's index (GI) for the scale insect populations in Figure 3.1. Values for the χ^2 test statistic [Eq. (3.7)] are given along with the critical chi-square table values at the 0.975 and 0.025 probability levels*

Plant	Dispersion	ID	χ^2	Critical Values for Chi-Square 0.975	0.025	GI
A	Random	0.87	7.8	2.7 –	19.0	0.00
B	Clumped	6.07	54.6		> 19.0	0.17
C	Uniform	0.07	0.6	< 2.7		−0.03

TABLE 3.2 *Properties of three indices of dispersion. The value of each index at maximum regularity, randomness, and maximum clumping are given. Modified from Elliott (1973)[a]*

Index	Value of Index at Maximum Uniformity	Randomness	Maximum Clumping
Index of dispersion, $s^2/\bar{x}$	0	1	n
Index of clumping, $s^2/\bar{x} - 1$	−1	0	$n-1$
Green's index, $(s^2/\bar{x}) - 1/(n-1)$	$-1/(n-1)$	0	1

[a]Key: $\bar{x}$ = mean number of individuals per SU, s^2 = variance, and n = total number of individuals in sample.

If $|d| < 1.96$, agreement with a Poisson series (a random dispersion) is accepted $(P > 0.05)$. If $d < -1.96$, a regular dispersion is suspected, and if $d > 1.96$, a clumped dispersion is likely (Elliott 1973). When N is large (> 30), a goodness-of-fit test to the Poisson distribution should also be conducted (see Section 3.2.1) and the results compared to the preceding variance-to-mean ratio test. Of course, both tests use the same data and are, therefore, not independent. Greig-Smith (1983) discusses situations where one of these two tests may detect obvious nonrandomness when the other fails to do so! This again points out the danger of basing conclusions on a single set of observations and a single test.

As seen from the preceding discussion, the variance-to-mean ratio (ID) is useful as a statistical test for assessing the agreement of a set of data to the Poisson series. However, in terms of measuring the degree of clumping, ID is not very useful. In Table 3.2 we show the values ID will have at maximum uniformity (i.e., each SU contains the same number of individuals), at random-

STATISTICAL ECOLOGY

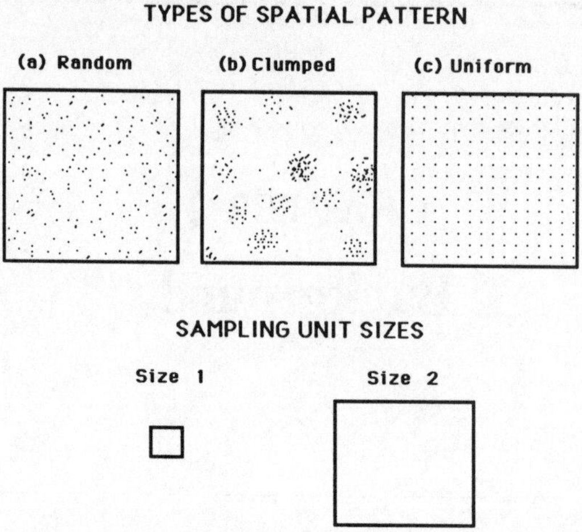

(a) Random   (b) Clumped   (c) Uniform

STEP 2. ESTIMATION OF PARAMETERS. The lognormal distribution is completely characterized by two parameters, S_0 and a [Eq. (7.1)]. We will illustrate some simple methods for estimating values for these parameters, but we alert the student to the fact that nonlinear regression could also be used (see Sokal and Rohlf 1981).

An approximation for parameter a is given by

$$a = \sqrt{\frac{\ln[S(0)/S(R_{max})]}{R^2_{max}}} \tag{7.3}$$

where $S(0)$ is the observed number of species in the modal octave and $S(R_{max})$ is the observed number of species in the octave *most distant* from the modal (indicated by R_{max}). In the example in Step 1, $R_{max} = 4$. Equation (7.3) works best when the data include observations at least three or four octaves distant from the mode. Also, if the most distant octave is the same in both directions, say -5 and $+5$, then a value for a using *both* values of $S(R_{max})$ in Eq. (7.3) should be computed and averaged.

Parameter a has been found to be about 0.2 for a large number of samples in ecology; consequently, this has lead to numerous hypotheses why this should be so. However, May (1975) has shown that this "rule" is actually a product of the mathematical properties of the lognormal distribution; as the total number of species in the community varies from 20 to 10,000, parameter a will vary from 0.3 to 0.13. (This assumes the underlying distribution follows what is called the "canonical lognormal" distribution—see Section 7.6.)

An estimate of parameter S_0 is obtained either by fixing it at the observed value for the number of species in the modal octave, $S(0)$, or by estimating it from

$$S_0 = e^{(\overline{\ln S(R)}+a^2\overline{R}^2)} \tag{7.4}$$

where $\overline{\ln S(R)}$ is the mean of the logarithms of the observed number of species per octave, a is estimated from Eq. (7.3), and $\overline{R}^2$ is the mean of the R^2s.

Traditionally, the lognormal distribution has been fitted by eye (Vandermeer 1981). However, Eq. (7.3) and (7.4) provide reasonable statistical estimates of the parameters. The program LOGNORM.BAS (on accompanying disk) allows an iterative substitution of different values for a and S_0 into Eq. (7.1) until the deviations between the observed (Step 1) and the expected number of species in the octaves (Step 3) are minimized.

STEP 3. THE EXPECTED FREQUENCIES. Using the estimates for S_0 and a, the expected lognormal frequencies are computed using Eq. (7.1) and the goodness of fit of the model to the observed frequencies is "tested" with a chi-square statistic. The degrees of freedom are equal to the number of octave classes minus two. Since we are only attempting to obtain an approximate fit, this chi-square statistic should be used as a guide for the selection of parameters rather than as a formal statistical test.

7.3 EXAMPLE: INSECTS ALONG A ROAD MEDIUM

The following insect count data were obtained from a grassland community within a road medium by combining the catches from 14 sweep nets

(X):	1	2	3	4	5	6	7	9	10	11	21	28	33	120
(f):	32	8	9	2	3	3	3	2	1	2	1	1	1	1

where X is the number of individuals per species and f is the frequency of species in each of the X classes. The total number of individuals equals 389 and the total number of species is 69.

Fitting a Lognormal Distribution

STEP 1. OBSERVED FREQUENCIES.   The number of species per octave are given in Table 7.1 and are plotted in Figure 7.3. Note that the second octave, which has 20 species, is the modal octave. Also, note how those species with abundance classes falling on the lines separating consecutive octaves (i.e., 1, 2, and 4) are split between those octaves:

Octave 1 (0–1): $32/2 = 16$

Octave 2 (1–2):   16 (from octave 1) + 8/2 = 20

TABLE 7.1   *The road medium insect count data arranged in octaves*

Octave	Number of Individuals per Species	R	R^2	Observed $S(R)$	$\ln S(R)$
1	0–1	−1	1	16	2.77
2	1–2	0	0	20	3.00
3	2–4	+1	1	14	2.64
4	4–8	+2	4	10	2.30
5	8–16	+3	9	5	1.61
6	16–32	+4	16	2	0.69
7	32–64	+5	25	1	0.00
8	64–128	+6	36	1	0.00

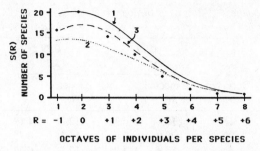

Figure 7.3 Frequency distribution of insect count data plotted in octaves: observed data shown as solid dots and fitted lognormal models as curves 1 ($a = 0.29, S_0 = 20$), 2 ($a = 0.29, S_0 = 13.2$), and 3 ($a = 0.305, S_0 = 15.9$).

Octave 3 (2–4):   4 (from octave 2) + 9 + 2/2 = 14

Octave 4 (4–8):   1 (from octave 3) + 3 + 3 + 3 + 3 + 0 = 10

STEP 2. ESTIMATION OF PARAMETERS.   An estimate of a is obtained from Eq. (7.3) with $R_{max} = 6$, $S_0 = 20$, $S(R_{max}) = 1$, and $R_{max}^2 = 36$ as

$$a = \sqrt{\frac{\ln(20/1)}{36}} = 0.29$$

From Eq. (7.4)

$$\overline{\ln S(R)} = (2.77 + \cdots + 0)/8 = 1.63$$
$$\bar{R}^2 = (1 + \cdots + 36)/8 = 11.5$$

and, thus

$$S_0 = e^{[1.63+(0.29)^2(11.5)]} = 13.2$$

The parameter estimates using the BASIC program LOGNORM.BAS are $a = 0.305$ and $S_0 = 15.9$. These results are summarized in Table 7.2.

STEP 3. EXPECTED FREQUENCIES.   Using $a = 0.29$ and $S_0 = 20$ and 13.2, the expected frequencies are computed using Eq. (7.1). For example, $S(R)$ for the first octave with $S_0 = 20$ is given by

$$S(-1) = 20e^{-(0.29)^2(-1)^2} = 18.4$$

TABLE 7.2 *Goodness-of-fit test for the lognormal model to observed insect count data. Expected results are given for three sets of parameter estimates (see text)*

Octave	R	Observed S(R)	$a =$ 0.29 $S_0 =$ 20 Expected S(R)	χ^2	$a =$ 0.29 $S_0 =$ 13.2 Expected S(R)	χ^2	$a =$ 0.305 $S_0 =$ 15.9 Expected S(R)	χ^2
1	−1	16	18.4	0.31	12.2	1.19	14.5	0.16
2	0	20	20.0	0.00	13.2	3.45	15.9	1.06
3	+1	14	18.4	1.05	12.2	0.27	14.5	0.02
4	+2	10	14.3	1.31	9.5	0.03	11.0	0.08
5	+3	5	9.5	2.10	6.3	0.25	6.9	0.52
6	+4	2	5.3	2.04	3.5	0.64	3.6	0.70
7	+5	1	2.5	0.90	1.7	0.26	1.6	0.20
8	+6	1	1.0	0.00	0.7	0.17	1.1	0.02
		69	89.4	7.71	59.3	6.26	69.1	2.76

Note that the sum of the expected frequencies for the three lognormal curves (given by the three sets of parameter estimates) are about 89, 59, and 69, respectively, yet the observed number of species is 69. Why is this? Recall that the total area under the lognormal curve represents the theoretical number of species available for observation, S^*. This can be computed from

$$S^* = 1.77(S_0/a) \tag{7.5}$$

For the three sets of parameter estimates given previously, $S^* = 122, 80$, and 92, respectively. This is more obvious when examining the plots of the three lognormal distributions in Figure 7.3: curve *1* has the most area under the curve; curve *2* has the least; and curve *3* is intermediate. Thus, returning to the expected frequencies in Table 7.2, we can see that the third set of parameter estimates (i.e., the iterative computations) is probably a good fit to these data and that there are about 23 species unobserved in the community. Of course, this is a simplified data set used for illustration, and large samples are important when fitting a lognormal.

7.4 EXAMPLE: COVER DATA FOR DESERT COMMUNITIES

For this example, we use Whittaker's (1965) data for species-rich desert communities found on the lower slopes of the Santa Catalina Mountains, Arizona. These data are shown in Figure 7.2. Whittaker used the percent cover of species as the measure of abundance and, arranged the data in octaves as:

R:	-6	-5	-4	-3	-2	-1	0	$+1$	$+2$	$+3$	$+4$	$+5$	$+6$
$S(R)$:	4	7	8	10	14	17	19	15	14	7	5	0	1

where $S(R)$ are the observed number of species in each octave, R octaves distant from the mode ($R = 0$). From Figure 7.2 it can be seen that the octave represented by $R = -6$ is the cover from 0.008 to 0.015 and so forth.

Using the program LOGNORM.BAS iteratively with initial values of $a = 0.247$ and $S_0 = 17.9$ [from Eq. (7.4)], the lowest chi-square (10.1) was obtained with $a = 0.235$ and $S_0 = 17.2$. Whittaker's (1965) estimates for a and S_0 were 0.245 and 17.5, respectively, obtained, presumably, by nonlinear regression. This lognormal curve is plotted in Figure 7.2. Note that the total number of observed species was 122 and, from Eq. (7.5), the theoretical number available for observation, S^*, is 129.

7.5 EXAMPLE: BREEDING BIRDS OF NEW YORK

This example is taken from Preston (1948), who summarized census data of breeding birds in a tract of land in western New York State. Below are data for the number of breeding pairs of each species (in octaves):

R:	-6	-5	-4	-3	-2	-1	0	$+1$	$+2$	$+3$	$+4$	$+5$
$S(R)$:	1	1.5	6.5	8	9	9	12	6	9	11	4	3

Again, $S(R)$ represents the number of species in the octave R from the modal ($R = 0$).

Using the program LOGNORM.BAS we computed best-fit estimates of $a = 0.21$ and $S_0 = 10$, with a chi-square = 7.05. These data and the lognormal curve are illustrated in Figure 7.4. Again, note that the value for a is near the

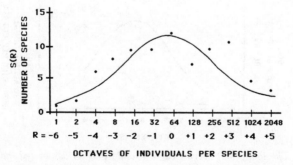

Figure 7.4 Breeding bird census data (dots) and a fitted lognormal distribution ($a = 0.21, S_0 = 10.0$).

often observed value of 0.2 (Step 2, Section 7.2). Whereas the observed number of species is 80, the theoretical number available for observation, S^*, is about 84 [Eq. (7.5)].

There are numerous data sets in Preston (1948) that can be used to gain familiarity with the program LOGNORM.BAS. Your estimates of a and S_0 obtained with LOGNORM.BAS can be compared to those given by Preston.

7.6 ADDITIONAL TOPICS IN DISTRIBUTION MODELS

While the lognormal tends to fit a large number of examples in ecology, there is some disagreement as to whether this reflects biological processes (e.g., community interactions) or is simply related to the statistical laws of large numbers (Minshall et al. 1985).

In Section 7.2 we discussed the interesting "rule" that values for parameter a of the lognormal distribution [Eq. (7.1)] are usually around 0.2. Another empirical rule associated with the lognormal distribution is the *canonical hypothesis* of Preston (1962). Preston analyzed a large set of community data with the lognormal model to examine the relationship between the number of species per octave and the total number of individuals in species per octave. An example of this type of comparison is illustrated in Figure 7.5. The "species curve" in Figure 7.5 is the number of species in each octave [$S(R)$, Eq. (7.1)], whereas the "individuals" curve is the number of individuals in each octave, $I(R)$, which is obtained from

$$I(R) = S(R)N(R) \qquad (7.6)$$

where $N(R)$ is the number of individuals per species in the Rth octave. Preston compared these two curves in terms of the relationship between R_{max}, the

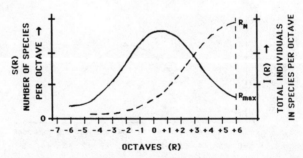

Figure 7.5 *The lognormal species curve (solid) and its associated individuals curve (dashed).*

octave for the most abundant species of the species curve, and R_N, the octave of the mode of the individuals curve. Their ratio

$$q = R_N/R_{max} \tag{7.7}$$

indicates that the lognormal species curve is *canonical* if $q = 1$, as illustrated in Figure 7.5. This empirical rule tends to apply to many community data that have been analyzed in this way.

May (1975) presented a thorough analysis of these two "rules" associated with the lognormal distribution. May showed that there exists a general relationship between the total number of species and individuals, S and N, and the parameters a and q. For values of S from 20 to 10,000 and for N from $10S$ to $10^7 S$, parameter a assumes values in the range 0.1–0.4 and q assumes values in the range 0.5–1.8. May (1975) suggests that this reflects statistical generalities associated with large samples.

However, Sugihara (1980) has demonstrated that the canonical hypothesis is obeyed far too strictly by ecological communities to be totally explained by May's (1975) statistical arguments. Sugihara plotted S from many communities against the standard deviation (σ) of the lognormal distribution. It turns out that there appears to be a unique relation between S and σ for a given value of q, and Sugihara found that natural communities closely approximate the canonical form (where $q = 1$). Sugihara proposed a biological mechanism that could give rise to this pattern.

Minshall et al. (1985) used the lognormal model as an empirical measure of the equilibrium state of invertebrate communities in river ecosystems. They suggested that a high degree of conformance with the lognormal model indicates that a community is in a high degree of equilibrium. Thus an evaluation of conformity to the lognormal across seasons could be used as a measure of the equilibrium state of the system. This is, of course, a good example of the use of a model for *pattern detection*, regardless of whether the underlying hypothesis proposed for the model is true or not. While this approach worked well in the Minshall et al. (1985) study, Gray (1981) found that a similar approach failed in an effort to detect stream pollution, because good fits to the lognormal occurred regardless of conditions and changes in community composition.

7.7   SUMMARY AND RECOMMENDATIONS

1. When the relative abundance of each species in a community is plotted in order of its rank from the most to least abundant, characteristic patterns emerge, as shown in Figure 7.1.

2. Ecologists have proposed a number of hypotheses in an attempt to explain the observed abundance relationships. However, contradictory hypotheses have been found to lead to the same model and, in some cases, different models may fit the same data. Thus, no general model has been found.

3. For large assemblages of species, relative abundance data tend to follow a lognormal distribution, perhaps only the result of the law of large numbers. However, most communities seem to conform to a special form of the lognormal model, the canonical. Thus, we recommend that the lognormal model be used to examine patterns of species abundance in communities.

4. What we have in the wake of all the proposed models and confusing jargon about species abundance relationships are some useful tools for detecting patterns or trends. These are useful in generating testable hypotheses about the organization of communities.

CHAPTER 8

Diversity Indices

In this chapter we describe several diversity indices that can be used to characterize species abundance relationships in communities. Diversity is composed of two distinct components: (1) the total number of species and (2) evenness (how the abundance data are distributed among the species). Since diversity indices often attempt to incorporate both of these components into a single numerical value, much debate and, often, confusion has resulted regarding their interpretation and correct usage. In this chapter we hope to clarify some of these issues.

8.1 GENERAL APPROACH

The concept of species diversity in community ecology has been intensely debated by ecologists over the years. In fact, Hurlbert (1971), went so far as to suggest that diversity was probably best described as a "nonconcept" because of the many semantic, conceptual, and technical problems associated with its use. In spite of debates and numerous cautionary remarks put forth by many regarding their use, diversity indices have remained very popular with ecologists. As is true for most methods, it is relatively simple to obtain some rudimentary knowledge and then forge ahead with computations; it is much more of a challenge to obtain a critical perspective. Hence, our first recommendation to the beginning student is to be aware of the limitations of all diversity indices.

Species diversity may be thought of as being composed of two components. The first is the number of species in the community, which ecologists often refer to as *species richness*. The second component is *species evenness* or equitability. *Evenness* refers to how the species abundances (e.g., the number of individuals, biomass, cover, etc.) are distributed among the species. For example, in a community composed of ten species, if 90% of the individuals belong to a single species and the remaining 10% are distributed among the nine other species, evenness would be considered low. On the other hand, if each of the ten species accounted for 10% of the total number of individuals, evenness would be considered maximum.

Over the years, a number of indices have been proposed for characterizing species richness and evenness. Such indices are termed *richness indices* and *evenness indices*. Indices that attempt to combine both species richness and evenness into a single value are what we refer to as *diversity indices*. The major criticism of all diversity indices is that they attempt to combine and, hence, confound a number of variables that characterize community structure: (1) the number of species, (2) relative species abundances (evenness), and (3) the homogeneity and size of the area sampled (see James and Rathbun 1981). These problems are considered below as we describe the procedures for computing diversity indices.

8.2 PROCEDURES

8.2.1 Richness Indices

It would appear that an unambiguous and straightforward index of species richness would be S, the total number of species in a community. However, since S depends on the sample size (and the time spent searching), it is limited as a comparative index (Yapp 1979). Hence, a number of indices have been proposed to measure species richness that are independent of the sample size. They are based on the relationship between S and the total number of individuals observed, n, which increases with increasing sample size.

Two historically well-known richness indices are as follows: Index 1, the Margalef (1958) index,

$$R1 = \frac{S - 1}{\ln(n)} \tag{8.1}$$

and Index 2, the Menhinick (1964) index,

$$R2 = \frac{S}{\sqrt{n}} \tag{8.2}$$

How useful are these indices? To answer this, consider the practical consequences of applying $R2$, the Menhinick index. Suppose $R2$ is used as a measure of species richness of avian communities at three sample locations along a mountain gradient. At the lowest elevation, let $S = 30$, $n = 100$ and, thus, $R2 = 3$. At the middle elevation, $S = 15$, $n = 25$ and, again, $R2 = 3$. Finally, at the highest elevation, $S = 10$, $n = 25$ and, now, $R2 = 2$. Based on these values for $R2$, should we conclude that species richness of the avian communities is greatest at the lower to mid-elevations on the mountain and that there are no differences until the highest elevation is reached? Before drawing any conclusions, it is important to recognize that the use of $R2$ presupposes that a functional relationship exists between S and n in the community and is, in fact, given by $S = k\sqrt{n}$, where k is constant. This must hold or $R2$ will vary with samples containing different values of n and, consequently, communities cannot be compared. Therefore, in the preceding example, conclusions regarding species richness are conditional on whether or not the functional relationship between S and n is actually given by $S = k\sqrt{n}$ and, if so, that k is constant. As Peet (1974) points out, if these assumptions fail to hold, the richness index will vary with sample size in some unknown manner. Consequently, we recommend using $R1$ and $R2$ as richness indices only if these assumptions are clearly met. We suspect that for most cases there will be no prior justification that these assumptions are satisfied and no conclusions should be made.

An alternative to richness indices is to use direct counts of species numbers in samples of *equal size*. Not only is this a very simple procedure, it also avoids some of the problems of using indices of the type described above. In situations where sample sizes are not equal (probably the usual situation), a statistical method known as *rarefraction* may be used to allow comparisons of species numbers between communities (Hurlbert 1971, Sanders 1968). To use the rarefraction method we assume that sample size biases or sampling differences between communities can be overcome by some underlying sampling model that applies to all communities concerned. We give an example of such a model below.

Hurlbert (1971) shows that the number of species that can be expected in a *sample* of n individuals [denoted by $E(S_n)$] drawn from a *population* of N total individuals distributed among S species is

$$E(S_n) = \sum_{i=1}^{S} \left\{ 1 - \left[\binom{N - n_i}{n} \bigg/ \binom{N}{n} \right] \right\} \qquad (8.3)$$

where n_i is the number of individuals of the ith species. In words, Eq. (8.3) computes the expected number of species in a random sample of size n as the sum of the probabilities that each species will be included in the sample.

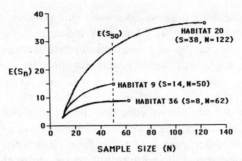

Figure 8.1 *Rarefraction curves for three avian habitats showing the expected number of species as a function of sample size. After James and Rathbum (1981).*

Details of how to actually use Eq. (8.3), which involves the computation of probabilities, are given in the computer program RAREFRAC.BAS (see accompanying disk); what we want to illustrate here is its potential as a richness measure.

An excellent example of the application of rarefraction can be found in James and Rathbun (1981), who studied 37 breeding bird censuses from a wide range of habitats in the United States and Canada. We highly recommend this paper to students interested in the details of this method. In Figure 8.1, a portion of some of their results are shown. The census conducted in habitat 20 (a hardwood-pine meadow) yielded a total of 38 species (*S*) from observing a total of 122 birds (*N*); this point is represented as an open circle in Figure 8.1. To compute the rarefraction curve for this habitat, they used Eq. (8.3) to compute the expected number of species at different sample sizes i.e., $E(S_n)$, at $n = 120$, 110, 100, etc. This is shown as the upper curve in Figure 8.1. (Note that *N* is treated here as a population parameter.) The same procedure was followed for habitat 9 (a jack pine-birch forest) with $S = 14$ and $N = 50$ and for habitat 36 (a mesquite–tamarisk–creosotebush desert) with $S = 8$ and $N = 62$. The rarefraction curves for each habitat can now be used to overcome their differences in total number of birds (i.e., $N = 122$, 50, and 62, respectively). James and Rathbun used a sample size of $n = 50$ as their standard, corresponding to the smallest sample size of all 37 habitats censused; this is illustrated as the vertical dashed line in Figure 8.1 at $n = 50$. At this sample size, we can order these three habitats in terms of their species richness. Habitat 20 has the highest richness, with an expected species number of 26.9, and habitat 36 has the lowest richness, with an expected species number of 7.8.

We agree with Hurlbert (1971) and James and Rathbun (1981) that rarefraction methods are preferred over the simple richness indices available when community sample sizes differ. However, Peet (1974) shows that for two communities possessing very different number of species and relative abundances, rarefraction may predict that both communities have the same number of species at small sample sizes. Thus, when using this method, it is assumed that the communities being studied do not differ in their species–individuals relationships (Peet 1974). Thus, while we recommend the use of rarefraction in species counts, we reemphasize what we mentioned earlier—be aware of the limitations of any diversity method.

8.2.2 Diversity Indices

Diversity indices incorporate both species richness and evenness into a single value. Because of this, Peet (1974) terms these *heterogeneity indices*. Probably the biggest obstacle to overcome in using diversity indices is interpreting what this single statistic actually means. For example, in some cases a given value of a diversity index may result from various combinations of species richness and evenness. In other words, the same diversity index value can be obtained for a community with low richness and high evenness as for a community with high richness and low evenness. It follows, then, that if you are given just the value of a diversity index, it is impossible to say what the relative importance is of species richness and evenness.

In spite of such problems, ecologists commonly employ diversity indices in their research, often ignoring some well-know problems in their application. The overview given here is intended to introduce some of the more popular diversity indices with an emphasis on interpreting (where possible) what they mean.

There are, literally, an infinite number of diversity indices (Peet 1974). The units of these indices differ greatly, making comparisons difficult and confusing, which adds to the interpretation problem. We believe that the series of *diversity numbers* presented by Hill (1973b) are probably the easiest to interpret ecologically.

In equation form, Hill's family of diversity numbers are

$$NA = \sum_{i=1}^{S} (p_i)^{1/(1-A)} \qquad (8.4)$$

where p_i is the proportion of individuals (or biomass, etc.) belonging to the ith species. The derivation of this equation is given in Hill (1973b). Hill shows that the 0th, 1st, and 2nd order of these diversity numbers [i.e., $A = 0$, 1, and 2 in Eq. (8.4)] coincide with three of the most important measures of diversity. (Note: see Hill 1973b for explanation of Eq. (8.4) when $A = 1$.) Hill's diversity numbers are

$$\text{NUMBER 0:} \quad N0 = S \tag{8.5a}$$

where S is the total number of species,

$$\text{NUMBER 1:} \quad N1 = e^{H'} \tag{8.5b}$$

where H' is Shannon's index (defined below), and

$$\text{NUMBER 2:} \quad N2 = 1/\lambda \tag{8.5c}$$

where λ is Simpson's index (defined below).

These diversity numbers, which are in units of number of species, measure what Hill calls the *effective number of species* present in a sample. This effective number of species is a measure of the degree to which proportional abundances are distributed among the species. Explicitly, N0 is the number of *all* species in the sample (regardless of their abundances), N2 is the number of *very abundant* species, and N1 measures the number of *abundant* species in the sample. (N1 will always be intermediate between N0 and N2.) In other words, the effective number of species is a measure of the number of species in the sample where each species is weighted by its abundance.

Hill's numbers differ only in their tendency to include or ignore the rarer species in the sample (Alatalo and Alatalo 1977). As an example, consider a sample of 11 species and 100 individuals where the abundances are distributed as 90, 1, 1, 1, 1, 1, 1, 1, 1, 1, 1. Obviously, since *one* species is very abundant, we would expect N2 to be near 1, which it is. (N2 = 1.23.) N0 is, of course, equal to 11 and N1 is 1.74, a value intermediate between N0 and N2.

As indices of diversity, Hill's numbers are intuitively appealing to ecologists (Peet 1974). While some vagueness remains in their interpretation (Hill 1973b), we find them much less confusing than other indices available. To reemphasize, the units in Hill's numbers are *species* and as the number increases: (1) less weight is placed on rare species (recall that N0, the lowest number, is the number of *all* species in the sample) and (2) lower values are obtained for N1 and N2, since they measure the number of *abundant* and *very abundant* species, respectively, in the sample.

In this chapter we limit our coverage of diversity indices to N1 and N2. As diversity indices, N1 and N2 are suitable for addressing "any question that a heterogeneity index can answer" (Peet 1974). Also, a number of the other popular diversity indices are merely variants of N1 and N2. In the computations given below, we exclusively use count data, that is, the number of individuals (n_i) of each ith species and the total number of individuals (n) sampled. However, as Hurlbert (1971) and Lyons (1981) have pointed out, most ecologists agree that the importance of a species in a community should be based on such quantities as biomass or productivity if such data are available. These quantities can be used in place of counts. However, it is important to recognize that when such substitutions are made, the usual methods for computing the variances for diversity measures are no longer appropriate. We refer the student to Lyons (1981) for details.

Two indices are needed to compute Hill's diversity numbers: Simpson's index, λ, and Shannon's index, H'. They are described below.

DIVERSITY INDEX 1. SIMPSON'S INDEX, λ.   Simpson (1949) proposed the first diversity index used in ecology as

$$\lambda = \sum_{i=1}^{S} p_i^2 \tag{8.6}$$

where p_i is the proportional abundance of the ith species, given by

$$p_i = \frac{n_i}{N}, \qquad i = 1, 2, 3, \dots, S$$

where n_i is the number of individuals of the ith species and N is the known total number of individuals for all S species in the population. Simpson's index, which varies from 0 to 1, gives the probability that two individuals drawn at random from a population belong to the same species. Simply stated, if the probability is high that both individuals belong to the same species, then the diversity of the community sample is low.

Equation (8.6) applies only to finite communities where all of the members have been counted i.e., $n = N$, where n is the total number of individuals in the *sample* and N is the total individuals in the *population*. Since we usually work with infinite populations where it is impossible to count all members, Simpson (1949) developed an unbiased estimator ($\hat{\lambda}$) for sampling from a infinite population as

$$\hat{\lambda} = \sum_{i=1}^{S} \frac{n_i(n_i - 1)}{n(n - 1)} \tag{8.7}$$

In practice, Eqs. (8.6) and (8.7) are approximately equal when we substitute n for N in Eq. 8.6.

The reciprocal of $\hat{\lambda}$ yields Hill's second diversity number N2 [see Eq. (8.5c)]. N2 is preferred over $\hat{\lambda}$ by numerous researchers on theoretical and practical grounds (see Alatalo and Alatalo 1977, Peet 1974, and Routledge 1979).

DIVERSITY INDEX 2. SHANNON'S INDEX, H'.   The Shannon index (H') has probably been the most widely used index in community ecology. It is based on information theory (Shannon and Weaver 1949) and is a measure of the average degree of "uncertainty" in predicting to what species an individual chosen at random from a collection of S species and N individuals will belong. This average uncertainty increases as the number of species increases and as the distribution of individuals among the species becomes even. Thus, H' has two properties that have made it a popular measure of species diversity: (1) $H' = 0$ if and only if there is one species in the sample, and (2) H' is maximum only when all S species are represented by the same number of individuals, that is, a perfectly even distribution of abundances.

The equation for the Shannon function, which uses natural logarithms (ln), is

$$H' = -\sum_{i=1}^{S^*} (p_i \ln p_i) \tag{8.8}$$

where H' is the average uncertainty per species in an infinite community made up of S^* species with known proportional abundances p_1, p_2, p_3, ..., p_{S^*}. S^* and the p_i's are *population parameters* and, in practice, H' is estimated from a sample as

$$\hat{H}' = -\sum_{i=1}^{S} \left[\left(\frac{n_i}{n}\right) \ln \left(\frac{n_i}{n}\right) \right] \tag{8.9}$$

where n_i is the number of individuals belonging to the ith of S species in the sample and n is the total number of individuals in the sample. Equation (8.9) is the most frequent form of the Shannon index used in the ecological literature. However, this estimator is biased because the total number of species in the community (S^*) will most likely be greater than the number of species observed in the sample (S). Fortunately, if n is large, this bias will be small.

DeJong (1975) demonstrated that H' is linearly related to the logarithm of the number of species in the sample, which is the form of the Shannon index given by Hill's N1 [Eq. (8.5b)], where the units are *species*. In other words, N1 gives the number of species that would, if each were equally common, produce the same H' as the sample. This statement is probably best appreciated by a simple numerical example. Given a community sample of three species represented by 100, 50, and 100 individuals, $H' = 1.05$ and N1 = 2.87. Thus, 2.87 species of *equal* abundance gives an $H' = 1.05$.

8.2.3 Evenness Indices

When all species in a sample are equally abundant, it seems intuitive that an evenness index should be maximum and decrease toward zero as the relative abundances of the species diverge away from evenness. Hurlbert (1971) noted that evenness indices have this property if they can be represented either as

$$V' = \frac{D}{D_{max}}$$ (8.10a)

or as

$$V = \frac{D - D_{min}}{D_{max} - D_{min}}$$ (8.10b)

where D is some observed diversity index and D_{min} and D_{max} are the minimum and maximum values, respectively, that D can obtain. We will refer back to these two equations in the discussion below.

In an attempt to quantify the evenness component of diversity, a number of indices have been proposed. We limit our presentation to five, each of which may be expressed as a ratio of Hill's numbers. For an in-depth treatment of evenness indices, we refer the student to a study by Alatalo (1981).

EVENNESS INDEX 1 (E1).   Probably the most common evenness index used by ecologists is

$$E1 = \frac{H'}{\ln(S)} = \frac{\ln(N1)}{\ln(N0)}$$ (8.11)

This is the familiar J' of Pielou (1975, 1977), which expresses H' relative to the maximum value that H' can obtain when all of the species in the sample are perfectly even with one individual per species (i.e., $\ln S$). Note that $E1$ is in the form of Eq. (8.10a).

EVENNESS INDEX 2 (E2).   Sheldon (1969) proposed an exponentiated form of $E1$ as an evenness index:

$$E2 = \frac{e^{H'}}{S} = \frac{N1}{N0}$$ (8.12)

EVENNESS INDEX 3 (E3).   If $E2$ is written in the form of Eq. (8.10b), that is, with the minimum subtracted, it becomes the evenness index proposed by Heip (1974):

$$E3 = \frac{e^{H'} - 1}{S - 1} = \frac{N1 - 1}{N0 - 1}$$ (8.13)

EVENNESS INDEX 4 (E4).   Hill (1973b) proposed the ratio of N2 to N1 as an index of evenness:

$$E4 = \frac{1/\lambda}{e^{H'}} = \frac{N2}{N1} \tag{8.14}$$

This is the ratio of the numbers of very abundant to abundant species. Recall from the discussion above that as the diversity of a community decreases, that is, as one species tends to dominate, both N1 and N2 will tend toward one. Under such conditions, $E4$ converges toward the value of one (Peet 1974).

EVENNESS INDEX 5 (E5).   If $E4$ is written in the form of Eq. (8.10b), it becomes

$$E5 = \frac{(1/\lambda) - 1}{e^{H'} - 1} = \frac{N2 - 1}{N1 - 1} \tag{8.15}$$

$E5$ is known as the *modified Hill's ratio*. Alatalo (1981) shows that $E5$ approaches zero as a single species becomes more and more dominant in a community (unlike $E4$, which approaches one). This is clearly a desirable property for an evenness index and is why $E5$ is preferred over $E4$ (Alatalo 1981).

An evenness index should be independent of the number of species in the sample. Intuitively, it would seem reasonable that, regardless of the number of species present, an evenness index should not change. Peet (1974) has shown that J' [$E1$, Eq. (8.11)] is strongly affected by species richness; the addition of one rare species to a sample that contains only a few species (low S) greatly changes the value of $E1$. This sensitivity is illustrated in Table 8.1, where a species represented by only one individual is added to a sample containing three well-represented species. $E2$ and $E3$ [Eqs. (8.12) and (8.13)], like $E1$, are very sensitive to species richness. In contrast, $E4$ and $E5$ are relatively unaffected by species richness.

TABLE 8.1 *Evenness indices computed for two samples. Sample 2 differs from 1 only by the addition of one individual of a new species (modified from Peet 1974 and Alatalo 1981)*

Sample	S	Individual Abundances	E1	E2	E3	E4	E5
1	3	500, 300, 200	0.94	0.93	0.90	0.94	0.91
2	4	500, 299, 200, 1	0.75	0.71	0.61	0.94	0.90

The actual number of species present in the community (S^*) should be known when using $E1$, $E2$, or $E3$. In practice, however, S^* is usually estimated by the number of species present in the sample, S. This usually leads to an underestimate of S^* and, hence, introduces a numerical bias in the estimate of the index (Pielou 1977). Peet (1975) noted that while this bias per se does not prevent an ecological interpretation of a general pattern, a high sensitivity to sample variation may render an index useless. On the other hand, $E4$ and $E5$ remain relatively constant with sampling variations (such as the occurrence of a rare species, Table 8.1) and hence tend to be independent of sample size. This is because $E4$ and $E5$ are computed as ratios where S is in both the numerator and denominator, thus effectively canceling the impact of the number of species in the sample.

The relative evenness of communities can also be compared by noting the steepness of their rarefraction curves (Section 8.2.1). Higher evenness is equated with a steeper rarefraction curve. In the example shown in Figure 8.1, habitat 20 has the highest evenness and habitat 36 the lowest.

8.3 EXAMPLE: LIZARD DIVERSITY IN A DESERT

Let us assume that an ecologist counting lizards in a 1-hectare area of desert found 6 species among 32 total individuals (Table 8.2). Using these data, we will illustrate the computations for the different species richness, diversity and evenness indices.

8.3.1 Richness Indices

RICHNESS INDEX 1 (R1). MARGALEF'S INDEX   [Eq. (8.1)]:

$$R1 = (6 - 1)/\ln(32) = 5/3.47 = 1.44$$

TABLE 8.2 *Number of individuals counted for each of 6 species of lizards in a 1x hectare plot*

Lizard Species	Number of Individuals
Cnemidophorus tesselatus	3
Cnemidophorus tigris	15
Crotophytus wislizenii	1
Holbrookia maculata	1
Phrynosoma cornutum	10
Scleoporus magister	2
	32

RICHNESS INDEX 2 (R2). MENHINICK'S INDEX [Eq. (8.2)]:

$$R2 = 6/\sqrt{32} = 6/5.66 = 1.06$$

8.3.2 Diversity Indices

DIVERSITY INDEX 1. Simpson's index [Eq. (8.7)], the probability that two individuals selected at random will belong to the same species, is

$$\lambda = [(3)(2) + (15)(14) + (1)(0) + (1)(0) + (10)(9) + (2)(1)]/[(32)(31)]$$
$$= (6 + 210 + 0 + 0 + 90 + 2)/992 = 308/992 = 0.31$$

Hill's second diversity number [Eq. (8.5c)], the number of very abundant species, is

$$N2 = 1/0.31 = 3.22$$

DIVERSITY INDEX 2. Shannon's index [Eq. (8.10)], the average degree of uncertainty in predicting what species an individual chosen at random from a sample will belong to, is

$$H' = -[(3/32)\ln(3/32) + (15/32)\ln(15/32) + (1/32)\ln(1/32)$$
$$+ (1/32)\ln(1/32) + (10/32)\ln(10/32) + (2/32)\ln(2/32)]$$
$$= -[(-0.223) + (-0.355) + (-0.107) + (-0.107) + (-0.364)$$
$$+ (-0.172)] = -(-1.33) = 1.33$$

Hill's first diversity number [Eq. (8.5b)], the number of abundant species, is

$$N1 = e^{1.33} = 3.78$$

As expected, N1 is intermediate between N0 (6, the total number of species) and N2 (3.22).

8.3.3 Evenness Indices

Using $S = 6$, the five evenness indices are computed as follows [see Eq. (8.11–8.15)]:

$$E1 = \ln(3.78)/\ln(6) = 1.33/1.79 = 0.74$$
$$E2 = 3.78/6 = 0.63$$
$$E3 = (3.78 - 1)/(6 - 1) = 2.78/5 = 0.56$$

$$E4 = 3.22/3.78 = 0.85$$
$$E5 = (3.22 - 1)/(3.78 - 1) = 2.22/2.78 = 0.80$$

8.4 EXAMPLE: FISH IN A FLORIDA ESTUARY

Catch data reported by Livingston (1976) in a study of seasonal fluctuation of fish in a north Florida estuary (Table 8.3) are used here to compare species

TABLE 8.3   *Fish catch data for a north Florida estuary taken from Livingston (1976). These are the results of 24-hour collections gathered by trawling during December 1972 and June 1973*

Fish Species	December	June
Dasyatis sabina	10	3
Ophichthus gomesi	3	0
Dorosoma petenense	1	1
Anchoa mitchilli	278	377
Arius felis	1	21
Hippocampus erectus	1	0
Eucinostomus gula	3	0
Lagodon rhomboides	1	0
Bairdiella chrysura	21	0
Cynoscion arenarius	19	30
Cynoscion nebulosus	7	0
Leiostomus xanthurus	18	238
Menticirrhus americanus	217	4
Micropogon undulatus	40	279
Orthopristis chrysoptera	0	1
Peprilus paru	1	0
Prionotus scitulus	35	0
Prionotus tribulus	25	0
Etropus crossotus	29	0
Symphurus plagiusa	50	3
Monacanthus hispidus	1	0
Bagre marinus	0	1
Porichthys porosissimus	0	4
Polydactylus octonemus	0	590
Paralichthys lethostigma	0	5
Trinectes maculatus	0	5
Total number of individuals $N =$ 761		1,562
Total number of species $S =$ 20		15

TABLE 8.4   *Species diversity indices for two fish sampling periods in a north Florida estuary*

Indices	December, 1972	June, 1973
Richness		
N0	20.00	15.00
R1	2.86	1.90
R2	0.72	0.38
Diversity		
λ	0.23	0.26
H'	1.91	1.54
N1	6.78	4.69
N2	4.39	3.90
Evenness		
E1	0.64	0.57
E2	0.34	0.31
E3	0.30	0.26
E4	0.65	0.83
E5	0.59	0.79

richness, diversity and evenness indices (Table 8.4). The BASIC program SPDIVERS.BAS (on accompanying disk) was used for the computations.

The richness indices R1 and R2 decrease from the December to the June catch. This reflects the decrease in number of species (N0 = 20 to N0 = 15) and the large increase in the total number of individuals found in June. These richness values are computed here for illustrative purposes; thus, we will not attempt further interpretation. Rather, we refer to the rarefraction curve for these data shown in Figure 8.2. Using a sample size of 761, the expected number of species for the December catch $E(S_{761})$ is 20 and for the June catch $E(S_{761})$ is 13. Hence, we would conclude from this comparison where differences in sample size occur that species richness decreased in the June catch.

All of the diversity indices declined from December to June (except λ, of course, since the value of λ increases with decreasing diversity). N1 and N2 are in *species* units and are the easiest to interpret. These numbers indicate an increase in dominance of fewer species in the June catch. This is fairly clear for the June catch where N2, a measure of *very abundant* species, is 3.9 and, in fact, four species account for 95% of the abundance (Table 8.3). N1, a measure of the *abundant* species, is 4.69 for June, which is always going to be a larger value than N2.

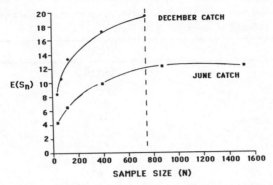

Figure 8.2   *Rarefraction curves for the expected number of fish species caught in December and June in a north Florida estuary. Data from Livingston (1976).*

The evenness indices are more difficult to interpret. We consider $E1$, $E2$, and $E3$ to be of limited value, because they are highly sensitive to the number of species in the sample. Both the Hill ratio ($E4$) and the modified Hill ratio ($E5$) increase from the December to the June sample. In this case, the increase seems to be related to the co-dominance by 4 species out of the 15 present and their very similar number of individuals (i.e., 377, 238, 279, and 590) compared to the remaining 11 species (Table 8.3).

8.5   EXAMPLE: FOREST TREE DIVERSITY

Zahl (1977) collected data on the total basal area of 20 trees in the Harvard research forest, Petersham, Massachusetts, during 1956, 1960, 1966, 1969, and 1975. His objective was to study patterns of changes in diversity with forest regeneration, since the forest was cut over in 1918 and then left undisturbed. For this example, we selected his data for 1956 and 1975 (Table 8.5).

Using the BASIC program SPDIVERS.BAS, species richness, diversity, and evenness indices were computed and are summarized in Table 8.6. These results lead to the same general conclusion as reached by Zahl (1977): there was a decrease in species richness, diversity, and evenness from 1956 to 1975. We recommend that the behavior of the indices for this example be closely compared with those of the previous example (Section 8.4), particularly with respect to the evenness indices.

TABLE 8.5   *Total basal areas at breast height in square meters for trees in the Harvard forest during 1956 and 1975. After Zahl (1977)*

Species	1956	1975
Quercus rubra	15.158	27.554
Betula papyrifera	8.601	11.845
Fraxinus americana	3.764	1.466
Acer rubrum	2.646	4.974
Pinus strobus	2.515	2.725
Betula lenta	2.240	2.210
Pinus resinosa	1.292	1.125
Hicoria ovata	0.622	0.680
Tsuga canadensis	0.527	0.808
Quercus alba	0.463	0.052
Acer saccharum	0.302	0.549
Quercus velutina	0.198	0.098
Pinus sylvestris	0.055	0.000
Betula lutea	0.024	0.030
Ostrya virginiana	0.012	0.009
Tilia americana	0.012	0.000
Populus grandidentata	0.006	0.000
Prunus serutina	0.000	0.012
Acer pennsylvanicum	0.000	0.021
Castanea dentata	0.000	0.009
Total basal area $N = 38.438$		54.166
Total number of species $S = 17$		17

8.6   ADDITIONAL TOPICS IN DIVERSITY

Peet (1974) classified diversity indices into two types based on their relative sensitivities to changes in the composition of a community: (1) type I—those most sensitive to changes in rare species, for example, $N1$; and (2) type II—those most sensitive to changes in common species, for example, $N2$. In Table 8.7, we give an example using artificial communities to illustrate these sensitivities.

Consider three communities, A, B, and C, each having a total of 100,000 individuals distributed among three common species and a variable number of rare species. Community A is composed of three common species and 200 rare species, each rare species having 1% of the total number of individuals (i.e., 200). When these rare species are changed to represent only 0.01% of the total number of individuals (i.e., 20, community B), the value of $N2$ is relatively unaffected while the value of $N1$ increases 151% (Table 8.7). This

TABLE 8.6   *Species diversity indices for two years in the Harvard forest based on the total basal area of 20 trees. These results are based on the data given in Table 8.5*

Indices	1956	1975
Richness		
N0	17.00	17.00
R1	4.38	4.01
R2	2.74	2.31
Diversity		
λ	0.21	0.31
H'	1.84	1.55
N1	6.27	4.71
N2	4.79	3.25
Evenness		
E1	0.65	0.55
E2	0.37	0.28
E3	0.33	0.23
E4	0.76	0.69
E5	0.72	0.61

TABLE 8.7   *A comparison of diversity indices showing their sensitivity to changes in rare and common species (Spp). Each artificial community has N = 100,000 individuals, three common species, and a variable number of rare species*

	Community					
	A		B		C	
Index	Spp.	Number of Individuals	Spp.	Number of Individuals	Spp.	Number of Individuals
	1	20,000	1	20,000	1	10,000
	2	20,000	2	20,000	2	10,000
	3	20,000	3	20,000	3	40,000
	4	200	4	20	4	20
	5	200	5	20	5	20
	⋮	⋮	⋮	⋮	⋮	⋮
	203	200	2,003	20	2,003	20
S	203			2,003		2,003
λ	0.12	— 0% →		0.12	— +50% →	0.18
H'	3.45	— +27% →		4.37	— −3% →	4.23
N1	31.55	— +151% →		79.24	— −15% →	68.99
N2	8.12	— +2% →		8.28	— −33% →	5.55

large change in N1 is typical for type I indices (Peet 1974). In community C, the relative amounts of the 3 common species were altered (leaving the rare species at 20 individuals each) and the indices recomputed. While both N1 and N2 changed, as expected, N2 (a type II index) was most affected, being reduced by 33%, while N1 was only reduced 15%.

A topic that also needs to be discussed concerns some problems associated with estimating diversity for an infinitely large population. A complete census is usually impossible and a sample must be taken, which is subject to error. Often the organisms to be studied occupy a continuum of space, so that some artificial sample unit (SU) must be used (e.g., a quadrat for a plant community or a trawl net for a fish community). Then, of course, the results are affected by such things as the spatial distributions of the populations and the size and shape of the SU. Zahl (1977) notes that the sampling distribution of any index of diversity will be a function of all of these factors.

Pielou (1966, 1975) proposed an interesting procedure for estimating diversity, and its sampling error, which attempted to account for some of these sampling problems. Briefly, Pielou's method, known as the *pooled-quadrat method*, involves taking q random quadrats (SUs) from a community, arranging these q quadrats in a random order, and then computing the diversity of the first quadrat, then the first plus second, and so on, each time pooling the observations of the quadrats. The value of the index computed for the accumulated quadrats will increase initially but tend to level off as the pooling continues. The value of the index after leveling off is then taken as the estimate of the diversity for the community.

Recently, several authors have developed alternatives to Pielou's method. Zahl (1977) introduced a jackknife method for estimating any of the diversity indices presented in this chapter. Zahl's method yields estimates of the standard deviations of these indices, which are necessary to test hypotheses and estimate confidence intervals (Heltshe and DiCanzio 1985). Heltshe and Forrester (1983a, 1985) show that under certain conditions, Zahl's jackknife estimator and Pielou's pooled-quadrat estimator are equivalent. Also related to this problem, Heltshe and Forrester (1983b) applied the jackknife approach to estimate S^*, the total number of species in a community. Since details of the jackknife method are beyond the scope of this book, we refer students to the preceding references.

The estimate of the total number of species in the community, S^* [from the lognormal model, Eq. (7.5)], has also been used as a richness index, but it requires a statistical fit of species abundance data to the lognormal frequency distribution, unlike Eq. (8.1) and (8.2). Although there is much more effort required in computing S^* (both in data requirements and computations), we would recommend S^* as a richness index over $R1$ and $R2$.

8.7 SUMMARY AND RECOMMENDATIONS

1. Be aware of the limitations of all diversity measures. This includes species richness indices, rarefraction models, diversity indices, and evenness indicies. These measures are easy to compute, but are usually difficult to interpret (Section 8.1).

2. The species richness indices $R1$ and $R2$ should not be used unless there is solid justification that their underlying functional relationships and assumptions, in fact, hold (Section 8.2.1). We suggest that, in practice, this will rarely be the case.

3. If sample sizes are equal, we recommend that direct counts of species be used to compare richness between communities. Where sample sizes differ, we recommend the use of the rarefraction model. However, it is important that underlying species–individuals relationships be similar (Section 8.2.1).

4. The use of S^* from the lognormal model is recommended as an index of species richness.

5. We recommend the use of Hill's diversity numbers $N1$ and $N2$ as measures of species diversity. They have been shown in several studies to be more interpretable than other diversity indices and have the appeal of being in units of species numbers (Section 8.2.2).

6. As a measure of evenness, we recommend the use of the modified Hill's ratio, $E5$, as it is least ambiguous and most interpretable. It does not require an estimate of the number of species in the community, which is affected by sample size (Section 8.2.3).

PART FOUR
SPECIES AFFINITY

CHAPTER 9

Background

Ecological communities are composed of a number of coexisting species. Some communities may have a large number of species (e.g., a tropical forest); others may have just a few (e.g., a polluted river). In Chapters 7 and 8 we described some empirical models for quantifying the relationships between the total number of species in a community and some measure of their abundances (e.g., total numbers). In this part of the book, we are interested in examining the affinities of coexisting species. How do coexisting species utilize common resources?

Consider, for example, a species-rich lake that has four dominant fish, all about the same size. Are they in direct competition for food and space? Do some species feed exclusively in the surface waters, while others feed on the lake bottom? When we spatially locate species A, are we likely more often than not to find species B there as well? In a broad sense, we can define such interspecific interactions as the degree of affinity between species.

One measure of affinity is the degree to which species *overlap* in their utilization of common resources. This overlap is defined in terms of various portions of the species niche that is shared by other species. Niche studies are based on such species attributes as diet, microhabitat preference, and timing of activities (e.g., foraging). Measures of niche overlap are presented in Chapter 10.

In Chapter 11 we cover the topic of interspecific association. In this instance, we are concerned only with measuring how often two species are found together in the same location. This affinity (or lack of it) for coexistence is tested by examining if the occurrence of the species [in a series of sample units (SUs)] is greater than or less than what would be expected if they were independent. If either positive or negative association is detected, we can measure the strength of this association with indices.

Association is based solely on presence/absence data. If a sample contains quantitative measures of species abundance, we can determine the covariation in abundances between species. This may lead to questions concerning species affinities. For example, if the abundance of one species always decreases when the other species increases, is there some type of causal negative interaction? Measures of interspecific covariation are presented in Chapter 12.

Each of these approaches is intended to help the ecologist detect patterns in species interactions. Of course, nothing about the underlying causes of a pattern can be inferred simply from its detection, although we hope that pattern detection will lead directly to testable hypotheses.

9.1 MATRIX VIEW

Cattell (1952) noted that the ecological data matrix could be studied from two distinct viewpoints: (1) down columns (the SUs) or (2) across rows (the species). Depending on which of these options is chosen, certain measures of *resemblance* are available. A taxonomy of these resemblance functions is given in Figure 9.1. It is important to recognize that the appropriate choice of

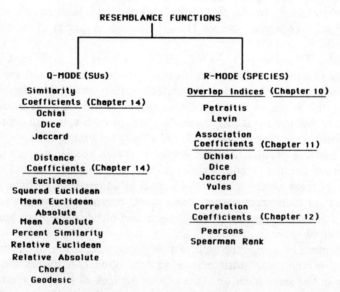

Figure 9.1   *Resemblance functions applicable to Q-mode analysis (similarity, distance) and R-mode analysis (overlap, association, correlation).*

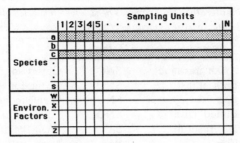

Figure 9.2 The shaded area indicates the form of the ecological data matrix for measuring species affinity. Interest is in the occurrences or abundances of species across sampling units.

a function is related to the fact that, in the ecological data matrix, we consider the species (rows) to be *dependent* on one another, whereas the SUs (columns) are *independent* samples (Legendre and Legendre 1983).

From our background discussion, the student will recognize that our interest in Chapters 10–12 is in species affinity, that is, measuring pairwise species resemblance based on data across rows in the ecological data matrix (Figure 9.2). Ecologists refer to this as *R*-mode analysis (Legendre and Legendre 1983).

The *R*-mode resemblance functions (Figure 9.1) are divided into overlap indices (Chapter 10), association coefficients (Chapter 11), and covariation coefficients (Chapter 12). These *R*-mode indices measure the dependence or intensity of the affinity between species. Intuitively, this makes sense because we are measuring the resemblance of species that occur together in SUs.

Q-mode resemblance functions measure the *similarity* or *dissimilarity* between SUs in terms of their species composition (i.e., down columns). Again, this terminology makes sense since we are comparing how similar or dissimilar *independent* samples are. Parts V through VII of this book are largely concerned with *Q*-mode analyses. *Q*-mode resemblance functions are presented in Chapter 14. A graphical representation of *R*- and *Q*-mode approaches is given in Figure 13.2.

Whereas the scheme presented in Figure 9.1 seems straightforward, there are many cases in the ecological literature where *R*- and *Q*-mode indices have been interchanged. For example, the association coefficients we describe in Chapter 11 have also been used to measure similarity between SUs (Chapter 14). Because of the nature of some of these particular coefficients, such usage is acceptable; however, using an *R*-mode coefficient like the correlation coefficient to measure *Q*-mode resemblance is not recommended (Orloci 1972, 1978).

TABLE 9.1 *Selected literature for examples of studies on species affinity involving niche overlap (SNO), interspecific association (SA), and interspecific covariation (SC)*

Location	Community	Method	Reference
South Dakota	Avian grassland	SNO	Garrat and Steinhorst 1976
Florida	Epibenthos	SNO	Livingston et al. 1976
St. Croix	Reef fish	SNO	Gladfelter and Johnson 1983
Caribbean	Butterfly fish	SNO	Findley and Findley 1985
California	Desert shrubs	SA	Went 1942
Lake Erie	Fish	SA	Nash 1950
Trinidad	Forest	SA	Greig-Smith 1952b
Colorado	Prairie	SA	Cook and Hurst 1963
Wisconsin	Forest herbs	SA	Smith and Cottam 1967
Florida	Leaf miners	SA	Bultman and Faeth 1985
England	Pastures	SC	Kershaw 1961
England	Chalk grassland	SC	Greig-Smith 1983
Utah	Zoobenthic	SC	Drake 1984

9.2   SELECTED LITERATURE

The measurement of niche overlap, interspecific associations and covariations have been done for a wide variety of biotic communities (Table 9.1).

CHAPTER 10

Niche Overlap Indices

How do coexisting species utilize common resources in a community? Species having similar patterns of resource usage may be thought of as having a high degree of "overlap"; those species with dissimilar usage patterns are considered to have low overlap. Ecologists have developed various indices that measure the degree of species overlap in an attempt to gain insight into community structure. In this chapter, we present the overlap measures of Petraitis (1979).

10.1 GENERAL APPROACH

Stemming from the classical work of Gause (1934), community ecologists have long sought to understand how coexisting species utilize common resources, such as food or space. Such research has led to theories on the natural regulation of species diversity in communities and to efforts aimed at refining the concept of the species niche (Schoener 1974). Recall that a species niche was defined in Chapter 7 as the "position" of a species in a community, including, for example, its use of food resources, time of activity, location, mode of interaction with other species, and so forth.

One of the early attempts to quantify the notion of species coexistence in terms of the niche was by Pianka (1973). Pianka wanted to measure how various coexisting lizard species compared with regard to their selection of food size and their location and time of foraging. He introduced an empirical index to measure the extent of similarity or *overlap* of the different species with regard to these resource states. In principle, two species with similar niche requirements would be expected to show a high degree of overlap. In theory, niche overlap is considered to be one of the possible determinants of species diversity and community structure (Petraitis 1979).

Niche overlap indices are now used extensively by ecologists and, not surprisingly, the literature abounds with a wide variety of proposed forms. In Chapter 9, we classified overlap indices as an R-mode approach. However, indices traditionally used in Q-mode studies have also been used as measures of overlap. In a broad sense, these overlap indices may be classified as distance measures (e.g., Levins 1968), association indices (e.g., Cody 1974), correlation coefficients (Pianka 1973), information measures (e.g., Horn 1966) and test statistics (e.g., van Belle and Ahmad 1974) (see Figure 9.1). Reviews on this subject can be found in Pielou (1972a), Abrams (1980), Hurlbert (1978), Lawlor (1980), and Zaret and Smith (1984).

Hurlbert (1978) recommended that the selection of a overlap index should be based on its ease of biological interpretation and its ability to account for variation in the availability of resources states. With regard to the latter, species may, for example, be compared on the basis of morphology, foraging time, food resources consumed, microhabitat use, etc. (Lawlor 1980); these are termed *resource states*. We give an example in Table 10.1 where four orders of insects are designated as resource states (i.e., prey) for three bird species (i.e., predators). Bird species 1 can be seen to feed evenly across these prey, whereas species 2 and 3 forage exclusively on prey items 1–2 and 3–4, respectively. However, when the relative abundance of the resource states is considered, a more informative interpretation emerges. Species 1 rejects the common resource items, that is, prey items 1 and 2 (which make up 80% of the total resource pool), to the point where it utilizes all resources equally. Bird species 2 feeds exclusively on abundant prey items 1 and 2 and, hence, would reject only one out of every five random encounters with potential prey in this environment. Bird species 3 would reject four out of every five, since

TABLE 10.1 *Hypothetical data showing four resource states (insect orders) as prey for three bird species. Modified from Petraitis (1979)*

		Insect Prey Items (Resource States)			
Insect Order:		(1)	(2)	(3)	(4)
Relative Availability of Resources:		0.4	0.4	0.1	0.1
	Species				
Relative Utilization by	(1)	0.25	0.25	0.25	0.25
Bird Species	(2)	0.50	0.50	0	0
	(3)	0	0	0.50	0.50

it prefers the least abundant prey. Intuitively, the student should sense that the "overlap" of resource utilization by bird species 1 with 2 is quite different than that of 1 with 3 when resource availabilities are taken into account.

Schoener (1974) and Hurlbert (1978) suggested weighting of the relative utilization of a resource by each species by the availability of that resource. However, most overlap indices are based on the relative usages of resource states without accounting for their relative availabilities (i.e., it is usually assumed that resources are equally available). Consequently, these indices must be used with great caution (Hurlbert 1978, Lawlor 1980, Petraitis 1979). Of course, it must be recognized that this type of availability data are rarely obtainable in field studies. Furthermore, our subjective decisions as to what constitutes "relative" availability for a species may or may not correspond to what a particular species actually perceives. Recognizing these limitations, Petraitis (1979) developed a measure of overlap based on the likelihood that the utilization of resources by one species is identical to that of another species. His *specific* and *general overlap indices* are presented below.

10.2 PROCEDURES

The relative use of resource states by each species is termed its *utilization curve*. For example, in Table 10.1, the utilization curve for bird species 3 is [0, 0, 0.5, 0.5] and for species 1 is [0.25, 0.25, 0.25, 0.25].

Levins (1968) proposed an index that measures the degree of overlap of two utilization curves. The Levins index for overlap (LO) of species 1 with species 2 is given by

$$LO_{1,2} = \frac{\sum_j^r [(p_{1j})(p_{2j})]}{\sum_j^r (p_{1j}^2)} \qquad (10.1)$$

where the terms are defined in Table 10.2. For our example in Table 10.1,

$$LO_{1,2} = \frac{(0.25)(0.50) + \cdots + (0.25)(0)}{0.25^2 + \cdots + 0.25^2} = 1.0$$

and, similarly, the overlap of species 2's utilization curve with species 1 is

$$LO_{2,1} = \frac{(0.25)(0.50) + \cdots + (0.25)(0)}{0.5^2 + \cdots + 0^2} = 0.5$$

Thus, the utilization curve of species 1 completely overlaps that of species 2, but the utilization curve of species 2 only overlaps one-half that of species 1.

TABLE 10.2 *Summary of terms for computing the specific and general measures of overlap. Three resource classes (r = 3) are used by two species (S = 2). Subscripts i, j represent the ith species and the jth resource. Modified from Petraitis (1985)*

	Species	Resource Classes (r)			
		(1)	(2)	(3)	Total
Tally	(1)	$n_{1,1}$	$n_{1,2}$	$n_{1,3}$	N_1
	(2)	$n_{2,1}$	$n_{2,2}$	$n_{2,3}$	N_2
Total Tally		t_1	t_2	t_3	T
Proportion	(1)	$p_{1,1}$	$p_{1,2}$	$p_{1,3}$	
	(2)	$p_{2,1}$	$p_{2,2}$	$p_{2,3}$	
Combined Proportions		c_1	c_2	c_3	

$$N_i = \sum_{j=1}^{r} (n_{ij}) \qquad t_j = \sum_{i=1}^{S} (n_{ij}) \qquad T = \sum_{j=1}^{r} (t_j)$$

$$p_{ij} = n_{ij}/N_i \qquad c_j = t_j/T$$

This difference is a result of how the amount of "overlap" between two species [i.e., the numerator in Eq. (10.1)] is standardized. The denominator in Eq. (10.1) is termed the *breadth* (*B*) of the *i*th species:

$$B_i = \frac{1}{\sum_j^r (p_{ij}^2)} \tag{10.2}$$

Note that the breadth of species 1 (from Table 10.1) is 4.0 and the breadth of species 2 is 2.0. Also, Eq. (10.2) may appear familiar as this was presented in Chapter 8 [Eq. (8.5c) and (8.6)].

Overlap as measured by the Levins index does not account for differences in resource availability. Hurlbert (1978) attempted to correct this by weighting niche overlap by the availabilities of the resources. Overlap between species 1 and 2 using Hurlbert's index (HO) is

$$HO_{1,2} = \sum_j^r \frac{(p_{1j})(p_{2j})}{c_j} \tag{10.3}$$

where c_j is the relative abundance of the *j*th resource (see Table 10.2). Again, using the data in Table 10.1, overlap as measured by HO is

$$HO_{1,2} = \frac{(0.25)(0.50)}{0.4} + \cdots + \frac{(0.25)(0)}{0.1} = 0.625$$

and, for the overlap of species 2's utilization curve with species 1 is

$$HO_{2,1} = \frac{(0.50)(0.25)}{0.4} + \cdots + \frac{(0)(0.25)}{0.1} = 0.625$$

In this case, overlap is standardized by resource availabilities and not by species breadth as in the Levins index. Therefore, $HO_{1,2} = HO_{2,1}$. Hurlbert (1982) noted that HO will be equal to 1.0 in cases where "both species utilize each resource state in proportion to its abundance".

10.2.1 Specific Overlap

Like the preceding LO and HO, specific overlap (SO) is also based on a comparison of utilization curves. However, unlike LO and HO, the amount of specific overlap by species i onto species k is the probability that the utilization curve of species i could have been drawn from species k's utilization curve. Petraitis derived the equations given subsequently by examining whether or not an observed species utilization of resource states could have been drawn randomly from the environmental resource spectrum. Using the type of species-resource data illustrated in Table 10.2, it is possible to compute the probability that species i's use of resources (i.e., n_{i1}, n_{i2}, n_{i3}), could have been drawn from species k's utilization curve (i.e., p_{k1}, p_{k2}, p_{k3}).

STEP 1. STATE HYPOTHESES. Two general types of hypotheses can be tested: (1) the null hypothesis that two species completely overlap, that is, their utilization curves are equal, and (2) the null hypothesis that the specific overlap by species i onto one species, k, is greater than the overlap of species i onto a second species, m.

STEP 2. COMPUTE SPECIFIC OVERLAP. For ease of presentation, specific overlap is shown for a pair of species, 1 and 2. All terms and their definitions are set out in Table 10.2. Specific niche overlap of species 1 onto species 2 and species 2 onto species 1 over r resource classes is given by

$$SO_{1,2} = e^{E_{1,2}} \tag{10.4}$$

and

$$SO_{2,1} = e^{E_{2,1}} \tag{10.5}$$

respectively, where

$$E_{1,2} = \sum_{j}^{r} (p_{1j}\ln p_{2j}) - \sum_{j}^{r} (p_{1j}\ln p_{1j}) \tag{10.6}$$

and

$$E_{2,1} = \sum_{j}^{r} (p_{2j} \ln p_{1j}) - \sum_{j}^{r} (p_{2j} \ln p_{2j}) \qquad (10.7)$$

Note that the quantity on the far right-hand side of Eq. (10.6) and (10.7) is the Shannon function [see Eq. (8.8)]. Also, be aware that computation of SO requires both species to utilize *all* resource classes, since if a species utilization of a resource is zero (i.e., $p_{i,j} = 0$), then $\ln(p_{ij})$ in Eq. (10.6) or (10.7) is undefined. Finally, SO ranges from 0 to 1.

STEP 3. COMPUTE TEST STATISTICS.   First, to test the null hypothesis that the specific overlap of species i onto k is complete, we compute the statistic

$$U_{i,k} = -2N_i \ln(\mathrm{SO}_{i,k}) \qquad (10.8)$$

where $U_{i,k}$ is distributed as chi-square with $r - 1$ degrees of freedom. Thus, for specific overlap of species 1 onto species 2, we have

$$U_{1,2} = -2N_1 \ln(\mathrm{SO}_{1,2})$$

and for species 2 onto species 1,

$$U_{2,1} = -2N_2 \ln(\mathrm{SO}_{2,1})$$

Second, to test the hypothesis that specific overlap by species i onto species k is greater than the overlap of species i onto species m, we compute the log-likelihood ratio, W, as

$$W = N_i \ln(\mathrm{SO}_{i,k}/\mathrm{SO}_{i,m}) \qquad (10.9)$$

If $W > 2$, we conclude that specific overlap by species i onto species k is greater than the overlap of species i onto species m.

10.2.2 General Overlap

The amount of general overlap between species in a community is defined by Petraitis as the probability that the utilization curves of all species was drawn from a "common" utilization curve. There is no limit on the number of species to be compared (recall that specific overlap was the probability that one species utilization curve could have been drawn from another's). General overlap between species is computed as a weighted average of species utilization curves as follows:

STEP 1. STATE HYPOTHESIS. The hypothesis to be tested is that there is complete overlap of the species (i.e., the utilization curves of the species could have been drawn from a common utilization curve).

STEP 2. COMPUTE GENERAL OVERLAP (GO). General overlap is given by

$$GO = e^E \tag{10.10}$$

where

$$E = \frac{\sum_i^s \sum_j^r [n_{ij}(\ln c_j - \ln p_{ij})]}{T} \tag{10.11}$$

(Again, refer to Table 10.2 for definition of terms.) Note, the summation in Eq. (10.11) is across all resource classes ($j = 1$ to r) for all species ($i = 1$ to S).

STEP 3. COMPUTE TEST STATISTIC. The statistic

$$V = -2T \ln GO \tag{10.12}$$

is distributed as a chi-square variate with $(S - 1)(r - 1)$ degrees of freedom. If V exceeds the critical value for chi-square at, say, $P = 0.05$, then the null hypothesis of complete overlap is rejected.

10.3   EXAMPLE: CALCULATIONS FOR BUMBLEBEE COEXISTENCE

In a study of niche differentiation in species of Colorado bumblebees (*Bombus* spp.), Pyke (1982) examined the relationship between bee proboscis length and the corolla length of flowers visited. For this example, the number of visits of four species of bees over the "resource spectrum" of four classes of corolla lengths is used (Table 10.3).

10.3.1   Calculating Specific Overlap

STEP 1. STATE HYPOTHESIS. We can test the hypothesis that the specific overlaps of any pair of bee species is "complete" versus the alternative of "some" or "none." Furthermore, we can test the hypothesis that the specific overlap by bee species i onto species k is greater than the overlap of species i onto species m. Overlap in the context of this example refers to the selection of flower size classes by the bees.

TABLE 10.3 Bumblebee (Bombus spp.) data for four species over four resource classes (corolla length in millimeters for flowers visited) and specific overlaps

Tally Species		Resource Classes				
		(1) (0–4 mm)	(2) (4–8 mm)	(3) (8–12 mm)	(4) (>12 mm)	Total
B. appositus	(1)	27	47	357	925	1,356
B. flavifrons	(2)	1,018	1,363	1,139	964	4,484
B. frigidus	(3)	333	638	145	13	1,129
B. occidentalis	(4)	155	84	70	51	360
Total Tally		1,533	2,132	1,711	1,953	7,329
Proportion						
B. appositus	(1)	0.020	0.035	0.263	0.682	
B. flavifrons	(2)	0.227	0.304	0.254	0.215	
B. frigidus	(3)	0.295	0.565	0.128	0.012	
B. occidentalis	(4)	0.431	0.233	0.194	0.142	
Combined Proportions		0.209	0.291	0.233	0.266	
Specific Overlaps ($SO_{i,k}$)						
Species		(1)	(2)	(3)	(4)	
(1)		1.0	0.510	0.059	0.359	
(2)		0.385	1.0	0.574	0.912	
(3)		0.107	0.736	1.0	0.736	
(4)		0.226	0.902	0.675	1.0	

STEP 2. COMPUTE SPECIFIC OVERLAP (SO). For illustrative purposes, the specific overlap between bee species 1 and 2 and between 2 and 1 is computed. From Table 10.3 and Eq. (10.6),

$$E_{1,2} = [(0.02\ln 0.227) + \cdots + (0.682\ln 0.215)]$$
$$- [(0.02\ln 0.02) + \cdots + (0.682\ln 0.682)]$$
$$= -1.48 - (-0.81) = -0.673$$

and from Eq. (10.7)

$$E_{2,1} = [(0.227\ln 0.02) + \cdots + (0.215\ln 0.682)]$$
$$- [(0.227\ln 0.227) + \cdots + (0.215\ln 0.125)]$$
$$= -2.33 - (-1.38) = -0.955$$

Thus, from Eq. (10.4),

$$SO_{1,2} = e^{-0.673} = 0.510$$

and from Eq. (10.5)

$$SO_{2,1} = e^{-0.955} = 0.385$$

The corresponding values using the Levins index are $LO_{1,2} = 0.427$ and $LO_{2,1} = 0.898$. These very different overlaps for *B. appositus* onto *B. flavifors* and vice versa when measured with LO result from the large difference in niche breadths [Eq. (10.2)] for the two species ($B_1 = 1.87$ and $B_2 = 3.13$). The SO values for these two species are, on the other hand, more similar, which will generally be the case, since no such standardization is used. The remaining specific overlap values for these data are given in Table 10.3.

STEP 3. COMPUTE TEST STATISTICS.   From Eq. (10.8)

$$U_{1,2} = (-2)(1,356)(-0.673) = 1,826$$

and

$$U_{2,1} = (-2)(4,484)(-0.955) = 8,564$$

Both are well above the critical value of 7.82 (3 df, $P = 0.05$), therefore, the hypothesis of complete overlap is rejected. However, this is a somewhat limited test as the alternatives are either "no" overlap or "some" overlap (i.e., the relative strength of overlap is only being tested against "complete" overlap).

Pyke (1982) measured the average proboscis length (in mm) for the bee species to see if a relationship existed between proboscis length and corolla lengths of flowers visited. Proboscis lengths of the four bee species in Table 10.3 can be ranked in order from the largest to smallest as $1 > 2 > 3 = 4$. Based on the number of visits to flowers of different corolla lengths, we might expect a niche separation between bees of different proboscis lengths. Using the log-likelihood ratio, W [Eq. (10.9)], we can test whether or not the overlap of one species onto another species is greater than its overlap onto yet another species. For example, we might test the following two hypotheses:

Case (1)—Overlap of species 1 (largest proboscis) onto species 2 (intermediate-size proboscis) is greater than the overlap of species 1 onto species 3 (smallest proboscis). The log-likelihood ratio $W = 1,356 \ln(0.51/0.059) = 2,925$, which is much larger than the critical value of 2. Hence, we accept this hypothesis.

Case (2)—Overlap of species 2 (intermediate proboscis) onto species 3 (smallest proboscis) is greater than the overlap of species 2 with species 4 (also the smallest). The log-likelihood ratio $W = 4,484 \ln(0.574/0.912) = -2,076$, so we reject this hypothesis.

10.3.2   Calculating General Overlap

STEP 1. STATE HYPOTHESIS.   The hypothesis to be tested is that there is complete overlap in the selection of flower length by the four bee species. Based on the results of specific overlap, we expect to reject this hypothesis.

STEP 2. COMPUTE GENERAL OVERLAP.   Using Eq. (10.11)

$$E = [27(\ln 0.209 - \ln 0.20) + 47(\ln 0.291 - \ln 0.035) + \cdots$$
$$+ 51(\ln 0.291 - \ln 0.142)]/7329$$
$$= -1240.3/7329 = -0.169$$

and, therefore, from Eq. (10.10) the estimate of GO (noted as $\widehat{GO}$) is obtained as

$$\widehat{GO} = e^{-0.169} = 0.844$$

STEP 3. COMPUTE TEST STATISTIC.   Using Eq. (10.12), the value of $V = (-2)(7,329)(-0.169) = 2,480$, which is much greater than the critical chi-square value of 16.92 (9 df, $P = 0.05$). Thus, we reject the hypothesis of complete overlap, although the overlap index ($\widehat{GO} = 0.84$) is large. As previously mentioned, the relative strength of overlap between the bee species is being tested against "complete" overlap.

10.4   EXAMPLE: BIRD DIET OVERLAP

Root (1967) examined the gut contents of three species of birds to examine their similarities in prey selection (Table 10.4). This example is used by Petraitis (1979) to demonstrate his specific and general overlap measures.

Output from our BASIC program SPOVRLAP.BAS (see accompanying disk) is given in Table 10.5. The general overlap between the three birds in their diet selection is high ($\widehat{GO} = 0.85$), but far from "complete," as $V = 211.2$ with the critical value of chi-square = 15.5 (8 df, $P = 0.05$). The specific overlap by Hutton's Vireo (species 3) onto the Blue-gray Gnatcatcher (species 1) is 0.767, while its overlap onto the Warbling Vireo (species 2) is 0.711. Using the log-likelihood ratio $W = 134 \ln(0.767/0.711) = 10.16$, we can conclude

TABLE 10.4   *Arthropods in the diets of three coexisting bird species. Data from Root (1967)*

Species*	(1) Hemiptera	(2) Coleoptera	(3) Lepidoptera	(4) Hymenoptera	(5) Other
	Resource (Insect Prey)				
(1)	103	93	20	40	31
(2)	23	31	132	14	13
(3)	16	40	33	30	15

*(1) = *Polioptila caerulea* = Blue-gray Gnatcatcher.
(2) = *Vireo gilvus* = Warbling Vireo.
(3) = *Vireo huttoni* = Hutton's Vireo.

TABLE 10.5   *Computation of general and specific overlap using bird data from Root (1967) and the BASIC program SPOVRLAP.BAS*

A. General Species Niche Overlap

Number of spp.	$\widehat{GO}$	G_{min}	G_{adj}	V	df
3	0.847	0.349	0.764	211.2	8

B. Specific (Pairwise) Niche Overlap

Spp.	Pair	Index and Test Statistics		
i	k	SO	U	df
1	2	0.494	404.4	4
2	1	0.359	436.1	4
1	3	0.768	151.3	4
3	1	0.767	71.0	4
2	3	0.713	144.4	4
3	2	0.711	91.5	4

that the Hutton's Vireo shows greater overlap in diet selection with the Blue-gray Gnatcatcher than with the Warbling Vireo.

10.5   ADDITIONAL TOPICS IN NICHE OVERLAP

To illustrate some interesting aspects of the behavior of general overlap, we have tabulated four different kinds of data sets in Table 10.6. For cases *A*, *B*, and *C*, two species share no resources in common, whereas in case *D* their

TABLE 10.6 *Data for the utilization of two resources by two species along with values for specific (SO), general ($\widehat{GO}$), minimum GO (GO_{min}) and adjusted GO (GO_{adj}) overlap*

		Resource Class			Indices			
Case	Species	(1)	(2)	N_i	SO	$\widehat{GO}$	GO_{min}	GO_{adj}
A	(1)	10	0	10	0	0.82	0.82	0.0
	(2)	0	190	190				
B	(1)	100	0	100	0	0.52	0.52	0.0
	(2)	0	100	100				
C	(1)	50	0	50	0	0.57	0.57	0.0
	(2)	0	150	150				
D	(1)	50	50	100	1	0.50	0.50	1.0
	(2)	50	50	100				

utilization curves are identical. Since zero values are not allowable when using many of the equations of this chapter, we arbitrarily set zeros to 1×10^{-7} and used the program SPOVRLAP.BAS to compute values of GO and SO. For cases $A-C$, specific overlap for species 1 onto species 2 (and vice versa) is zero and, for case D, specific overlap is equal to one, as expected. However, in spite of the fact that there is no overlap in resource use between species 1 and 2 in cases $A-C$, general overlap ($\widehat{GO}$) varies from a high of 0.82 to a low of 0.50 (Table 10.6). Smith (1984) noted that there is a dependence of the lower bound of GO on sample size and the number of species considered. It turns out that the minimum value for GO can be computed based on the N_i's and T (Smith 1984) as

$$GO_{min} = e^{(1/T)[\sum_{i=1}^{s} (N_i \ln N_i) - (T \ln T)]} \tag{10.13}$$

Thus, GO_{min} is 0.82 in case A and 0.50 in case B. Smith (1984) recommended that estimates of general overlap, $\widehat{GO}$, should be adjusted to account for this dependence on sample size. One possibility is to scale the value of $\widehat{GO}$ in reference to its possible range as affected by GO_{min} as

$$GO_{adj} = \frac{\widehat{GO} - GO_{min}}{1 - GO_{min}} \tag{10.14}$$

For the examples in Table 10.6, values for GO_{adj} in cases of no overlap $(A-C)$ are now 0 and, in the case of complete overlap (D), GO_{adj} is 1. GO_{adj} will always range from 0 to 1, whereas $\widehat{GO}$ will vary anywhere from 0.5 to 1 in the two-species case. For more than two species, the lower bound of GO will decrease (Smith 1984). Note, however, that for statistical tests, the unadjusted value of $\widehat{GO}$ should always be used (Petraitis 1985).

In reading the ecological literature the student will come across numerous indices of overlap, usually named after the original authors (e.g., Levins, Pianka, Morisita, Horn, etc.). A limitation of most overlap measures is the difficulty in determining whether the degree of overlap is statistically significant (Zaret and Smith 1984). The question that often arises is at what value has an overlap index departed sufficiently from zero (no overlap) to indicate a significant degree of overlap? The specific and general indices developed by Petraitis (1979, 1985) are one approach to overcome this problem. Garratt and Steinhorst (1976) described a technique for testing the significance of Morisita's index and for defining confidence intervals based on the statistical method of nonparametric permutation tests. Zaret and Smith worked out analytical methods to compute test statistics and confidence intervals for several overlap measures, including $\widehat{GO}$. Finally, Mueller and Altenberg (1985) used jackknife and bootstrap methods for estimating variance and bias in Morisita's and Horn's indices of overlap.

10.6 SUMMARY AND RECOMMENDATIONS

1. Niche overlap indices are extensively used by ecologists. However, for the most part, their usefulness is limited, because they do not take resource availability into account (all resources are assumed to be equally available) and no statistical tests are available (Section 10.1).

2. With regard to resource availabilities, it is rare that such data are obtainable in field studies. Furthermore, what we may perceive as "available" may or may not correspond with what the species of interest sees. As is always true in ecological studies, our interpretation and manipulation of data must be done with a thoughtful eye on how we are limited by our initial powers of perception as to what are "important" data.

3. Petraitis (1979) introduced a measure of specific overlap (SO), which is based on the probability that the utilization curve (i.e., the proportional usage of each resource) of one species could have been drawn from another species' utilization curve (Section 10.2.1).

4. Statistical tests are available for testing the null hypothesis that the specific overlap of one species onto another is complete. However, if we reject this hypothesis, the alternatives are "none" or "some" overlap, thus somewhat limiting the usefulness of this test [Eq. (10.8)].

5. The log-likelihood ratio is a useful statistic that allows us to test the hypothesis that the overlap of one species onto another species is greater than its overlap onto yet another species [Eq. (10.9)].

6. The general measure of overlap, GO, developed by Petraitis (1979) is based on the likelihood that two or more species utilization curves are drawn from a "common" utilization curve (Section 10.2.2).

7. The minimum value that GO can assume is a function of sample size and the number of species. Thus, we recommend that GO be adjusted to account for its possible range for a particular set of data. We provide one possible form of adjustment in Eq. (10.14), so that GO_{adj} ranges from 0 to 1.

8. Both the specific (SO) and general (GO) measures of overlap require that all species utilize all resources (i.e., no zeros in the data), because of the use of logarithms in their computation. Strictly speaking, this does limit the usefulness of these indices as the potential certainty exists for zeros to occur in ecological data sets.

9. The interpretation of overlap indices must be done with caution, particularly in reference to intensity of competition. As Lawlor (1980) and Zaret and Rand (1971) point out, there may be no interpretable relation between some measure of low overlap and, say, competition intensity, if, for example, the low overlap is a result of "past" competition. We recommend that students read Lawlor (1980).

CHAPTER 11

Interspecific Association

Species interactions are of central importance in the ecology of a species. Within any given community, there are a number of biotic and abiotic factors that influence the distribution, the abundance, and, subsequently, the interactions of species. Depending on whether or not two species select or avoid the same habitat, have some mutual attraction or repulsion, or have no interaction whatsoever, a certain pattern of *interspecific association* results. This association may be positive, negative, or absent. In this chapter we describe methods for detecting the existence of association between species and present indices for measuring the *degree* of association. These techniques are based solely on the presence or absence of species in sampling units (SUs).

11.1 GENERAL APPROACH

We are concerned here with measuring how often two species are found in the same location. This affinity (or lack of it) for coexistence of two species is referred to as *interspecific association*. In general, an association between two species exists because: (1) both species select or avoid the same habitat or habitat factors; (2) they have the same general abiotic and biotic environmental requirements; or (3) one or both of the species has an affinity for the other, either attraction or repulsion (Hubalek 1982).

The detection of species association has important ecological implications. Some ecological processes that may result in a positive or negative association between two species are summarized in Table 11.1. Note, in particular, that it is possible for positive or negative association to exist even in the absence of

125

TABLE 11.1   *Ecological processes and interspecific interactions that may result in positive and negative association among species. Adapted from Schluter (1984)*

Interaction	Process Example	
	Negative	Positive
None	Species have different resource requirements	Species have a common response to a supply of unlimited resources
Mutualism	Resources compete and are used exclusively by species	Species enhance each other's survival probabilities
Competition	Interference between species produces occasional exclusion	Species fluctuate in unison in response to limited resources
Predation	High predator densities produce a local depression of prey	Predators fluctuate in positive response to variations in prey

interspecific interactions per se. Also, note that the outcome "no association" is not explicitly included in this list, since this may result from a balancing of negative and positive forces (Schluter 1984). As emphasized throughout this text, the detection of a pattern (interspecific association, for example) does not provide a causal understanding of why such a pattern might exist. Rather, pattern detection should ideally lead to the generation of hypotheses of possible underlying causal factors, which subsidary studies can be designed to address.

The study of species association involves two distinct components. The first is a statistical test of the hypothesis that two species are associated or not at some predetermined probability level. The second is a measure of the degree or strength of the association (Figure 9.1). These should be regarded as separate characteristics of an association. Below, we (1) outline a procedure used to test for association between pairs of species, (2) present three measures of interspecific association, and (3) describe a simultaneous test for detecting association in a large number of species.

11.2 PROCEDURES

The procedure for studying interspecific association is based on the presence or absence of species in a collection of SUs. We can represent this with binary data, that is, presence is indicated with a 1 and absence with a 0. The SUs may be either natural (e.g., feathers, decaying logs, leaves) or artificial (e.g., plots, quadrats, lines). (Recall that in Chapter 2 we discussed numerous problems of using artificial SUs in spatial pattern analysis; the same problems exist for association.) Depending on the size and shape of the SU, it is possible to influence the outcome of association. This dependence can be lessened if the selection of the SU is made relative to the size, shape, and spatial distribution of the species under study. The SU must be large enough to potentially include at least one individual of each species and yet not so large that one of these species is included in every SU (Greig-Smith 1983).

11.2.1   Test of Association (Two-Species Case)

STEP 1. DATA SUMMARY.   For each pair of species, A and B, we obtain the following:

a = the number of SUs where both species occur
b = the number of SUs where species A occurs, but not B
c = the number of SUs where species B occurs, but not A
d = the number of SUs where neither A nor B are found
N = the total number of SUs ($N = a + b + c + d$)

This information is conveniently summarized in the form of a 2×2 table (Figure 11.1). Both the test and measures of association presented below are based on these data.

The expected frequency of occurrence of species A in the SUs, which we will represent as $f(A)$, is given by

$$f(A) = \frac{a + b}{N}$$ (11.1a)

and, for species B, by

$$f(B) = \frac{a + c}{N}$$ (11.1b)

We assume that both species have occurred in at least one SU in the collection, that is, $f(A)$ and $f(B)$ are greater than 0.

		Species B		
		present	absent	
Species A	present	a	b	m=a+b
	absent	c	d	n=c+d
		r=a+c	s=b+d	N=a+b+c+d

Figure 11.1 2×2 *contingency or species association table.*

STEP 2. STATE HYPOTHESIS. The null hypothesis is that the species are independent (i.e., there is no association).

STEP 3. COMPUTE TEST STATISTIC. The 2×2 table contains *observed* values for each of the cells (a, b, c, and d) from the sample of size N. To test for association, we compute what the *expected* values for each cell would be if the occurrences of species A and B are, in fact, independent and compare them to the observed values. A chi-square test statistic can be used to test the null hypothesis of independence in the 2×2 table. The chi-square test statistic is computed as

$$\chi_t^2 = \sum \frac{(\text{observed} - \text{expected})^2}{\text{expected}} \tag{11.2}$$

which is a summation over the four cells of the 2×2 table.

The expected value for cell a is given by

$$E(a) = \frac{(a + b)(a + c)}{N} = \frac{rm}{N} \tag{11.3}$$

or, from Eq. (11.1),

$$E(a) = f(B)(a + b) = f(A)(a + c) \tag{11.4}$$

In words, Eq. (11.4) states that of the total number of SUs where species A was present (i.e., $a + b$), we expect that if A and B were independent, species B should also be present in proportion to its overall frequency in the SUs, that is, $f(B)$; and vice versa for species A's presence in SUs where species B is present.

Similarly, the expected values for cells b, c, and d are, respectively,

$$E(b) = \frac{ms}{N}, \quad E(c) = \frac{rn}{N}, \quad \text{and} \quad E(d) = \frac{sn}{N} \tag{11.5}$$

The chi-square test statistic [Eq. (11.2)] is now given as

$$\chi_t^2 = \frac{[a - E(a)]^2}{E(a)} + \cdots + \frac{[d - E(d)]^2}{E(d)} \tag{11.6}$$

A mathematically equivalent, but certainly simpler equation, which may be used instead of Eq. (11.6), is

$$\chi_t^2 = \frac{N(ad - bc)^2}{mnrs}$$ (11.7)

Actually, in addition to being simpler to use, since Eq. (11.7) does not require the computation of expected values, nor the differences between observed and expected values, the associated rounding errors are avoided.

The significance of the chi-square test statistic is determined by comparing it to the theoretical chi-square distribution. The 2 × 2 contingency table has one degree of freedom, since a contingency table with r rows and c columns has $(r - 1)$ times $(c - 1)$ degrees of freedom (Zar 1974). The theoretical chi-square value for 1 df at the 5% probability level is 3.84. If $\chi_t^2 > 3.84$, we reject the null hypothesis that the co-occurrence of species A and B is independent and conclude that they are associated.

There are two types of associations:

1. Positive—if observed $a > E(a)$, that is, the pair of species occurred together more often than expected if independent.
2. Negative—if observed $a < E(a)$, that is, the pair of species occurred together less often than expected if independent.

This comparison of observed a to $E(a)$, that is,

$$a - E(a) = (ad - bc)/N$$ (11.8)

results in the quantity $ad - bc$ appearing in the numerator of all χ^2-like formulations, such as Eq. (11.7).

If any cell in the 2 × 2 table has an expected frequency < 1 or if more than two of the table cells have expected frequencies < 5, then the resulting chi-square test statistic will be biased (Zar 1974). A corrected chi-square is used to avoid biased values resulting from low cell expectations. In such cases, a continuity correction is applied to ensure a closer approximation to the theoretical, continuous chi-square distribution. This is achieved by using Yates's correction formula:

$$\chi_t^2 = \frac{N[|(ad) - (bc)| - (N/2)]^2}{mnrs}$$ (11.9)

11.2.2 Measures of Association (Two-Species Case)

Hubalek (1982) reviewed the properties of 43 indices that have been used to measure the degree of association between pairs of species. To sort through this plethora of indices, Hubalek identified five "admission" conditions. Indices that failed to satisfy any one of these conditions were deemed inadmissible and dropped from further consideration. The remaining admissible indices were then compared against eight optional criteria in order to help select the best association indices. Janson and Vegelius (1981) conducted a similar study where the characteristics of 20 association indices were examined over six "admission" conditions. The details of all these admission conditions are beyond the scope of our presentation, but five important conditions are listed here.

CONDITION 1.   Each association index should reach its minimum value at $a = 0$, that is, when the two species are never found together.

CONDITION 2.   The maximum value of the index should be when both species always occur together, that is, when $b = c = 0$.

CONDITION 3.   The association index should be symmetric, that is, the value of the index should be the same regardless of which species is designated "A" or "B" (Figure 11.1).

CONDITION 4.   The index should be able to discriminate between positive and negative associations. Formally, this means that the value of the index when $a > E(a)$ is always greater than when $a < E(a)$.

CONDITION 5.   The index should be independent of d, that is, the number of joint absences. There has been much debate as to whether the joint absence of species has any ecological meaning (Clifford and Stephenson 1975, Goodall 1978b, Sneath and Sokal 1973). We agree with Hubalek (1982) that indices using values of d are limited in ecology. For example, in a study of leaf miners on oak leaves by Bultman and Faeth (1985), average values for their 2×2 tables used to test for association were $a = 24$, $b = 875$, $c = 1,140$, and $d = 134,650!$ Any index using d would be "swamped" by the magnitude of d (joint absences).

Hubalek (1982) found six association measures to satisfy his admission conditions, and Janson and Vegelius (1981) found three that generally performed well. Three measures recommended by both studies—the Ochiai, Dice, and Jaccard indices—are presented below. These indices are equal to 0 at "no association" and 1 at "maximum association." The Ochiai and Dice measures are means of the ratios a/m and a/r, that is, the number of joint occurrences of the two species compared to the total occurrences of species A and B, respectively (Figure 11.1).

OCHIAI INDEX (OI).   The Ochiai (1957) index is based on the geometric mean of a/m and a/r, that is,

$$OI = \frac{a}{\sqrt{a+b}\sqrt{a+c}} \qquad (11.10)$$

DICE INDEX (DI). The Dice (1945) index is based on the harmonic mean of a/m and a/r, that is,

$$DI = \frac{2a}{2a+b+c} \qquad (11.11)$$

JACCARD INDEX (JI). This index is the proportion of the number of SUs where both species occur to the total number of SUs where at least one of the species is found:

$$JI = \frac{a}{a+b+c} \qquad (11.12)$$

To determine the sampling properties of a number of association measures, Goodall (1973) took repeated samples from a population with known species frequencies (a, b, c, and d) and computed the mean and variance of each index. Jaccard's index was found to be generally unbiased, even at small ($N = 10$) sample sizes. The Dice index tended to underestimate the true population values at small samples, but performed well at $N = 20$. Goodall did not test the Ochiai index.

11.2.3   Interspecific Association (Multiple-Species Case)

Usually, the association of more than a single pair of species is of interest; we may be interested in from 5, to perhaps 50 or more species. The number of all possible pairwise species associations or combinations that may be computed increases rapidly according to the equation $S(S - 1)/2$, where S is the number of species. For example, with five species there are $5(4)/2 = 10$ combinations; for 10 species, there are $10(9)/2 = 45$. Obviously, there is the problem of representing all the pairwise association index values in such a way as to ease interpretation. There are two ways of diagramming these multiple-species associations.

DIAGRAM 1. SPECIES ASSOCIATION COMPARISON MATRIX. All possible pair combinations of species associations can be displayed in a matrix of the form shown in Figure 11.2. To aid in interpretation, the species positions in the matrix can be reordered in such a way as to place species with highly significant positive index values along the diagonal of the matrix.

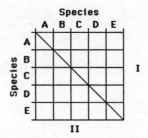

Figure 11.2 *Matrix display of all pairwise species associations. If two indices are computed, it is convenient to use the upper right (I) for one and the lower left (II) for the other.*

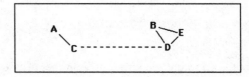

Figure 11.3 *Plexus diagram showing hypothetical associations between five species (A–E).*

DIAGRAM 2. *SPECIES CONSTELLATION (PLEXUS) DIAGRAMS.* A two-dimensional figure or *plexus diagram* (McIntosh 1978) can be used to summarize the pairwise associations found within a community. Plexus diagrams are constructed such that those species with positive associations are positioned close together and those with negative associations are positioned far apart. For example, if species *A* and *C* are positively associated as are species *B*, *D*, and *E*, and if species *C* has a weak association with species *D*, a plexus diagram could appear as shown in Figure 11.3.

The distances between species could be scaled in proportion to the magnitude of an association index. For example, the dimensions of the distances between species could be scaled between 0 and 1 to reflect the values of OI, DI, or JI. Since these diagrams are drawn by trial and error, as the number of species increases, it may require much effort to minimize distortions.

The test of association described in Section 11.2.1 is adequate for a single pair of species. However, this pairwise technique is inadequate when $S > 2$. Although we could compute all pairwise combinations of species association, they would not be independent and, consequently, we could not assign a probability to the distribution of outcomes (Schluter 1984). Pielou (1974) discusses the use of a supercritical chi-square value for this purpose. However, an

TABLE 11.2   Example of the presence (1) or absence (0) of S species ($i = 1, 2, 3, \ldots, S$) in N SUs ($j = 1, 2, 3, \ldots, N$)

Species	(1)	(2)	(3)	$\cdots$	(N)	Species Totals
	\multicolumn{5}{c}{SUs}	
(1)	1	0	1		0	n_1
(2)	1	0	1		1	n_2
(3)	0	1	0		0	n_3
$\vdots$	$\vdots$	$\vdots$	$\vdots$		$\vdots$	$\vdots$
(S)	0	0	1		1	n_S
SU Totals	T_1	T_2	T_3	$\cdots$	T_N	

alternative to testing the significance of many pairwise species associations would be to consider the significance of associations among many S species taken simultaneously. Pielou (1972b) suggests using a "2 to the Sth" contingency table, but this approach becomes impractical as the number of species increases. Schluter (1984) proposed a new approach using a variance ratio (VR) derived from a *null association model* to test simultaneously for significant associations. The VR index of association is easily derived from presence–absence data, and Schluter (1984) provides a statistic (W) to test for significant departures from the expected value of no association, where W approximates a chi-square distribution. Schluter's VR test is described below.

STEP 1. DATA SUMMARY.   An example of a data matrix showing the presence or absence of S species in N SUs is given in Table 11.2.

STEP 2. STATE HYPOTHESIS.   The null hypothesis is that there is no association among the S species. This is true under two conditions: (1) the species are independent, and (2) positive and negative associations between species cancel each other out (Schluter 1984). Thus, the alternative hypothesis is that there is a "net" positive or negative association among species.

STEP 3. COMPUTE TEST STATISTIC.   First, we compute the total sample variance for the occurrences of the S species in the sample as

$$\sigma_T^2 = \sum_{i=1}^{S} p_i(1 - p_i) \tag{11.13}$$

where $p_i = n_i/N$ (refer to Table 11.2 for definition of terms). Next, we estimate the variance in total species number as

$$S_T^2 = \frac{1}{N} \sum_{j=1}^{N} (T_j - t)^2 \qquad (11.14)$$

where t is the mean number of species per sample.

The variance ratio,

$$\mathrm{VR} = S_T^2/\sigma_T^2 \qquad (11.15)$$

serves as an index of overall species association. The expected value under the null hypothesis of independence is 1. VR > 1 suggests that, overall, the species exhibit a positive association. If VR < 1, a negative net association is suggested.

A statistic, W, is computed that may be used to test whether deviations from 1 are significant. For example, if the species are not associated, then there is a 90% probability that W lies between limits given by the chi-square distribution:

$$\chi_{.05,N}^2 < W < \chi_{.95,N}^2 \qquad (11.16a)$$

where

$$W = (N)(\mathrm{VR}) \qquad (11.16b)$$

Schluter (1984) noted that there may be situations where some species occur positively among themselves, but negatively with other species, yet the variance ratio test detects no association. Of course, even in a large random assemblage of species, there will be some significant associations detected using the pairwise chi-square procedure in Section 11.2.1; therefore, the variance ratio test is appropriate for large collections of species. Also, this test will often detect significant associations when pairwise tests do not.

11.3   EXAMPLE: CALCULATIONS

In Table 11.3, abundance and presence–absence data for three species from five SUs are given. In reality one would *never* attempt to analyze a data matrix based on such a small sample size, but for our purpose of providing a guide through the necessary computations for species association, these data are illustrative.

TABLE 11.3   *Ecological data matrix of (a) the abundance or (b) the presence–absence of three species (Spp) in five sampling units (SUs)*

(a) Abundance

Spp	(1)	(2)	(3)	(4)	(5)	Mean	Variance
	\multicolumn{5}{c}{SUs}		
(1)	2	5	5	3	0	3.0	4.5
(2)	0	3	4	2	1	2.0	2.5
(3)	2	0	1	0	2	1.0	1.0

(b) Presence–Absence

Spp	(1)	(2)	(3)	(4)	(5)	Totals (n_i)
	\multicolumn{5}{c}{SUs}	
(1)	1	1	1	1	0	4
(2)	0	1	1	1	1	4
(3)	1	0	1	0	1	3
Totals (T_j)	2	2	3	2	2	11

11.3.1 Test of Association (Two-Species Case)

STEP 1. DATA SUMMARY.   The 2×2 contingency table for species pair 1 and 3 is (with similar computations for other pairs)

Species 3

		present	absent	
	present	$a = 2$	$b = 2$	$m = 4$
Species 1	absent	$c = 1$	$d = 0$	$n = 1$
		$r = 3$	$s = 2$	$N = 5$

STEP 2. STATE HYPOTHESIS: "The occurrence of species 1 and species 3 in the five SUs is independent."

STEP 3. COMPUTE TEST STATISTIC.   From Eq. (11.3), we note that $E(a) = (4)(3)/5 = 2.4$, which is close to the observed value of $a = 2$. From Eq. (11.7),

$$\chi_t^2 = 5[(2)(0) - (2)(1)]^2/(4)(1)(3)(2) = 0.83,$$

and with continuity correction [Eq. (11.9)]

$$\chi_t^2 = 5[|(2)(0) - (2)(1)| - (5/2)]^2/(4)(1)(3)(2) = 0.052$$

Note that the continuity correction factor has a big effect on the value of the chi-square statistic, as it should when the sample sizes are small (as in the case here). Since the critical value of chi-square at 1 df is 3.84, we do not reject the null hypothesis, although we again remind the student that this example is only illustrative.

11.3.2   Measure of Association

Since species 1 and 3 were found to be independent (not associated), a measure of the strength of association is meaningless. However, for the purpose of illustrating computations, we compute the Ochiai, Dice, and Jaccard indices from Eq. (11.10)–(11.12):

$$OI_{1,3} = \frac{2}{\sqrt{2 + 2}\sqrt{2 + 1}} = 0.58$$

$$DI_{1,3} = \frac{(2)(2)}{(2)(2) + 2 + 1} = 0.57$$

$$JI_{1,3} = \frac{2}{2 + 2 + 1} = 0.40$$

11.3.3   Interspecific Association (Multiple-Species Case)

STEP 1.   The data are summarized in Table 11.3*b*.

STEP 2.   The hypothesis is that there is no association among the three species.

STEP 3.   The VR test statistic is computed by first calculating the total sample variance for the occurrences of the species in the samples as [Eq. (11.13)]:

$$\sigma_T^2 = (4/5)(1 - 4/5) + \cdots + (3/5)(1 - 3/5) = 0.56$$

The estimate of variance in total species number [Eq. (11.14)] is [where $t = (2 + 2 + 3 + 2 + 2)/5 = 2.2$]

$$S_T^2 = (1/5)[(2 - 2.2)^2 + \cdots + (2 - 2.2)^2] = 0.16$$

Thus, the variance ratio is [Eq. (11.15)]

$$VR = \frac{0.16}{0.56} = 0.28$$

which suggests a net negative association among the species. In fact, for each species pair, the observed $a < \exp(a)$. To test this deviation from 1, we compute [from Eq. (11.16b)]

$$W = (5)(0.28) = 1.43$$

Under the hypothesis of no association, there is a 90% probability that W should lie between the limits [Eq. (11.16a)]

$$1.14 < W < 11.07$$

Thus, we accept the null hypothesis of no association.

TABLE 11.4 *Ecological data matrix of (a) the abundance or (b) the presence–absence of five cockroach species from six Panamanian localities (modified from Wolda et al. 1983)*

(a) Abundance

| | | Locality | | | | | | |
| | | BCI | LC | FORT | BOQ | MIR | CORG | |
Species	No.	(1)	(2)	(3)	(4)	(5)	(6)	Mean
Ceuthobiella spp.	(1)	0	0	0	0	0	1	0.17
Compsodes cucullatus	(2)	14	38	1	1	4	0	9.67
Compsodes delicatulus	(3)	28	4	1	0	1	0	5.83
Buboblatta armata	(4)	7	0	0	0	0	0	1.17
Latindia dohrniana	(5)	68	29	0	0	11	24	22.00

(b) Presence–Absence

| | | Locality | | | | | |
| | | BCI | LC | FORT | BOQ | MIR | CORG |
Species	No.	(1)	(2)	(3)	(4)	(5)	(6)
Ceuthobiella spp.	(1)	0	0	0	0	0	1
Compsodes cucullatus	(2)	1	1	1	1	1	0
Compsodes delicatulus	(3)	1	1	1	0	1	0
Buboblatta armata	(4)	1	0	0	0	0	0
Latindia dohrniana	(5)	1	1	0	0	1	1

11.4   EXAMPLE: PANAMANIAN COCKROACHES

Blacklight light-trap sampling units were used by Wolda et al. (1983) to study species of *Blattaria* cockroaches from six localities in the Republic of Panama: Barro Colorado Island (BCI), Las Cumbres (LC), Fortuna (FORT), Boquet (BOQ), Miramar (MIR), and Corriente Grande (CORG). Presence–absence data for five cockroach species of the *Polyphagidae* are given in Table 11.4*b*. Using the BASIC program SPASSOC.BAS (see accompanying disk), chi-square and the interspecific association indices (OI, DI, and JI) were computed for all pairwise combinations of species, as well as the variance ratio test for multiple species associations (Table 11.5).

The variance ratio is 1.06, essentially equal to the expected value of 1.0 under the null hypothesis of no associations. The test statistic $W = 6.38$ falls within the range 1.63–12.6 [Eq. (11.16)] so we accept the null hypothesis.

There is only one species pair (*Ceuthobiella* spp. and *Compsodes cucullatus*) that in fact show a chi-square test statistic greater than 3.84; however, as expected, these values are all biased owing to the small sample size. (Note that the program SPASSOC.BAS checks for biased values.) If we had looked for

TABLE 11.5   *Interspecific association indices and test statistics between five* Polyphagidae *cockroaches in six Panamanian sites*

Multiple Species

VR, Index of overall association = 1.06
W, Test statistic = 6.39

Species Pairs

Species Pair	Association Type[a]	Chi-Square Value	Bias	Yates's Chi-square	Ochiai	Dice	Jaccard
1 2	−	6.00	*	0.96	0.00	0.00	0.00
1 3	−	2.40	*	0.15	0.00	0.00	0.00
1 4	−	0.24	*	0.96	0.00	0.00	0.00
1 5	+	0.60	*	0.15	0.50	0.40	0.25
2 3	+	2.40	*	0.15	0.89	0.89	0.80
2 4	+	0.24	*	0.96	0.45	0.33	0.20
2 5	−	0.60	*	0.15	0.67	0.67	0.50
3 4	+	0.60	*	0.15	0.50	0.40	0.25
3 5	+	0.38	*	0.09	0.75	0.75	0.60
4 5	+	0.60	*	0.15	0.50	0.40	0.25

[a] Sign indicates direction of the species associations.
* Uncorrected chi-square value is biased because either (1) expected frequency in any cell of the 2 × 2 table < 1 and/or (2) expected frequency of more than two cells < 5.

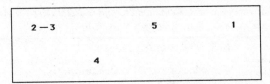

Figure 11.4 *Plexus diagram for the association of five cockroach species from Panamanian sites. Species connected by a solid line are positively associated.*

species affinities using only the association indices, we might have been tempted to conclude that, for example, species 2 and 3 are highly associated (Figure 11.4). Of course, the indices measure the degree of association and do *not* provide a test for association as chi-square does. This might suggest that additional studies using a larger sample are warranted.

TABLE 11.6 *Ecological data matrix of (a) abundances and (b) presence–absences for eight trees in 10 upland forest sampling units, southern Wisconsin (see Peet and Loucks 1977)*

(a) Abundances

Species		Sampling Units									
Name	No.	(1)	(2)	(3)	(4)	(5)	(6)	(7)	(8)	(9)	(10)
Bur oak	(1)	9	8	3	5	6	0	5	0	0	0
Black oak	(2)	8	9	8	7	0	0	0	0	0	0
White oak	(3)	5	4	9	9	7	7	4	6	0	2
Red oak	(4)	3	4	0	6	9	8	7	6	4	3
American elm	(5)	2	2	4	5	6	0	5	0	2	5
Basswood	(6)	0	0	0	0	2	7	6	6	7	6
Ironwood	(7)	0	0	0	0	0	0	7	4	6	5
Sugar maple	(8)	0	0	0	0	0	5	4	8	8	9

(b) Presence–Absence

Species		Sampling Units									
Name	No.	(1)	(2)	(3)	(4)	(5)	(6)	(7)	(8)	(9)	(10)
Bur oak	(1)	1	1	1	1	1	0	1	0	0	0
Black oak	(2)	1	1	1	1	0	0	0	0	0	0
White oak	(3)	1	1	1	1	1	1	1	1	0	1
Red oak	(4)	1	1	0	1	1	1	1	1	1	1
American elm	(5)	1	1	1	1	1	0	1	0	1	1
Basswood	(6)	0	0	0	0	1	1	1	1	1	1
Ironwood	(7)	0	0	0	0	0	0	1	1	1	1
Sugar maple	(8)	0	0	0	0	0	1	1	1	1	1

TABLE 11.7 *Interspecific association indices and test statistics for eight trees in ten upland forest sampling units, southern Wisconsin*

Multiple Species

VR, Index of overall association = 0.45
W, Test statistic = 4.45

Species Pairs

Species Pair		Association Type[a]	Chi-Square		Yates's Chi-square	Association Indices		
			Value	Bias		Ochiai	Dice	Jaccard
1	2	+	4.44	*	2.10	0.82	0.80	0.67
1	3	+	1.67	*	0.05	0.82	0.80	0.67
1	4	−	0.74	*	0.05	0.68	0.67	0.50
1	5	+	3.75	*	1.28	0.87	0.86	0.75
1	6	−	4.44	*	2.10	0.33	0.33	0.20
1	7	−	3.40	*	1.41	0.20	0.20	0.11
1	8	−	6.67	*	3.75	0.18	0.18	0.10
2	3	+	0.74	*	0.05	0.67	0.61	0.44
2	4	−	1.67	*	0.04	0.50	0.46	0.30
2	5	+	1.67	*	0.23	0.71	0.67	0.50
2	6	−	10.00	*	6.27	0.00	0.00	0.00
2	7	−	4.44	*	2.10	0.00	0.00	0.00
2	8	−	6.67	*	3.75	0.00	0.00	0.00
3	4	−	0.12	*	1.98	0.89	0.89	0.80
3	5	−	0.28	*	0.63	0.82	0.82	0.70
3	6	−	0.74	*	0.04	0.68	0.67	0.50
3	7	−	1.67	*	0.04	0.50	0.46	0.30
3	8	−	1.11	*	0.00	0.60	0.57	0.40
4	5	−	0.28	*	0.63	0.82	0.82	0.70
4	6	+	1.67	*	0.05	0.82	0.80	0.67
4	7	+	0.74	*	0.05	0.67	0.62	0.44
4	8	+	1.11	*	0.00	0.74	0.71	0.56
5	6	−	1.67	*	0.23	0.58	0.57	0.40
5	7	−	0.10	*	0.23	0.53	0.50	0.33
5	8	−	2.50	*	0.63	0.47	0.46	0.30
6	7	+	4.44	*	2.10	0.82	0.80	0.67
6	8	+	6.67	*	3.75	0.91	0.91	0.83
7	8	+	6.67	*	3.75	0.89	0.89	0.80

[a] Sign indicates direction of the species association.
* Uncorrected chi-square value is biased because either (1) expected frequency in any cell of the 2 × 2 table < 1 and/or (2) expected frequency of more than two cells < 5.

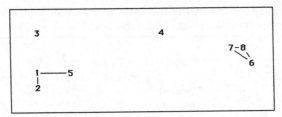

Figure 11.5 *Plexus diagram showing pattern of association between eight tree species in upland forest, southern Wisconsin. Species connected by solid lines are positively associated.*

11.5 EXAMPLE: WISCONSIN FORESTS

For this example, presence–absence data for eight tree species from 10 upland forest plots (SUs) in southern Wisconsin are used (Table 11.6). Using the BASIC program SPASSOC.BAS, the variance ratio test, chi-squares, and association indices were determined (Table 11.7). An overall negative association is suggested (VR = 0.44). However, $W = 4.45$ and falls within the range 3.96–18.3, so we accept the null hypothesis of no association among the eight species. If we look at the pairwise comparisons, we find that all of the computed chi-square values are biased. There does seem to be some strong possibilities that true negative associations (species pairs 1–8, 2–6, and 2–8) and positive associations (species pairs 6–8 and 7–8) might exist, as shown in a plexus diagram for the eight species (Figure 11.5). These results suggest that further studies might be warranted to confirm or refute these possible associations.

11.6 ADDITIONAL TOPICS IN INTERSPECIFIC ASSOCIATION

Pielou (1977) distinguished between *complete* and *absolute association*. Complete association occurs where one species is always present with the second but the second species may occur in some SUs without the first (Case 1, Table 11.8). This occurs when the second species is more common (i.e., cell b or c equal zero, but not both). Absolute association exists when neither species is ever found without the other (i.e., $b = c = 0$; Case 2, Table 11.8). It would be desirable for an index to be able to distinguish between absolute and complete associations, which, as can be seen in Table 11.8, the Ochiai, Dice, and Jaccard indices do.

The Ochiai, Dice, and Jaccard indices are examples of nonprobabilistic association indices. Many probabilistic (contingency table, chi-square-like) indices have also been proposed to measure the degree to which the quantity

TABLE 11.8 *Values of four measures of association for complete association*
(Case 1), absolute association (Case 2) and no association (Case 3)

Case (1)

		Species B present	Species B absent	
Species A	present	25	25	50
	absent	0	25	25
		25	50	75

Case (2)

		Species B present	Species B absent	
Species A	present	25	0	25
	absent	0	25	25
		25	25	50

Case (3)

		Species B present	Species B absent	
Species A	present	0	25	25
	absent	25	25	50
		25	50	75

Index	Case (1)	Case (2)	Case (3)
Ochiai	0.71	1.0	0.0
Dice	0.67	1.0	0.0
Jaccard	0.50	1.0	0.0
Yule	0.50	1.0	−0.5

$a - E(a)$ varies from 0 (reviewed in Hubalek 1982). As we discussed earlier
in this chapter, there are two distinct components of an association: (1) the
statistical tests of significance (using a chi-square test statistic) and (2) indices
to measure the degree of association. An example of a probabilistic index
that has been used to address *both* components of association is Yule's (1912)
index (YI):

$$YI = \sqrt{\frac{\chi_t^2}{N}} \qquad (11.17)$$

where χ_t^2 is the chi-square value as computed in Eq. (11.7). Yule's index ranges from -1 (when $a = d = 0$) to $+1$ (when $b = c = 0$), and is often referred to as the point correlation coefficient for discrete-binary (presence–absence) data (Pielou 1977). It can be seen from Table 11.8 that YI does distinguish between complete and absolute association. However, this index, unlike the Ochiai, Dice, and Jaccard indices, includes values for cell d in its computation, and while the two species in Case 3 (Table 11.8) do not occur together ($a = 0$), the value of YI $= -0.50$.

There are numerous theoretical and practical limitations to YI as a measure of the strength of an association (see Janson and Vegelius 1981, and Hubalek 1982). However, in practice, Hubalek found YI to work quite satisfactorily compared to other indices, and it does offer an interesting addition to the use of OI, DI, and JI.

Earlier we discussed Schluter's (1984) variance ratio test for detecting species associations based on presence–absence data. Schluter also extended his method to the use of density data for species. McCulloch (1985) gives connections between Schluter's association tests and some standard statistical tests.

11.7   SUMMARY AND RECOMMENDATIONS

1. One test of association between a pair of species is whether or not their joint occurrences over a series of SUs is independent. If they occur together more often than expected, they are positively associated; if their joint occurrences are less than what would be expected if independent, they are negatively associated.

2. There are two distinct components for the determination of a species association. First, there is a statistical test of species independence, and, second, there is a measure of the degree of strength of the association. We recommend that studies of interspecific association should include both components (Sections 11.2.1 and 11.2.2).

3. Although natural and artificial SUs may be used to obtain species presence–absence data, the methods presented in this chapter are best applied to natural SUs. Artificial SUs present numerous problems in evaluating the effect of the size and shape of the SU on species frequency.

4. There is a bewildering number of association indices in the ecological literature. We recommend the Ochiai, Dice, and Jaccard indices as ones that are simple to use and understand. Furthermore, based on extensive studies of the performance of association indices, these three indices have been judged to have excellent properties (Section 11.2.2.).

5. When the objective is to assess the presence or absence of association in a large group of species simultaneously, we recommend the variance ratio test of Schluter (Section 11.2.3).

6. Establishing the existence or absence of an association tells us nothing about the possible causes. We recommend using large sample sizes; the student should always exercise caution in interpreting results.

CHAPTER 12

Interspecific Covariation

When a sample contains quantitative measures of species abundance (e.g., biomass), the *covariation* in abundance between species can be assessed. If the abundances of two species tend to increase and decrease together, are they responding in the same way to the same environmental factors? If the abundance pattern for one species always increases when the other decreases, is there some possible negative interaction? The use of correlation coefficients to measure the relative intensity of covariation in pairwise species abundance data is presented in this chapter.

12.1 GENERAL APPROACH

In Chapter 11, we examined species association by testing for independence in the co-occurrence of pairs of species. This was based entirely on presence–absence data. If their co-occurrence was not independent, the species were determined to be either positively or negatively associated. If abundance data for species are available (e.g., density, percent cover, biomass), additional questions concerning species affinity may be posed. If the abundance of one species is high, does this somehow reduce, or perhaps enhance, the abundance of another species? This type of covariation question is entirely different from those posed in Chapter 11. In fact, two species may exhibit a strong positive association with regard to their joint occurrence in sampling unit (SUs), yet may have a strong negative covariation (i.e., when one species abundance increases, the other's decreases). For this reason, it is important to make a sharp distinction between species association and species covariation, an

145

argument previously made by Hurlbert (1969). Unfortunately, these two terms are too often inappropriately interchanged. The term *correlation* is general, but in many ecological contexts it should be reserved to imply (and measure) *how two species covary*.

In this chapter, the abundance patterns of species pairs in a series of SUs will be examined with correlation coefficients. The SUs may be either natural or artificial units, although, in the latter instance, caution must be used. The discussion concerning the use of artificial SUs in Section 11.2 applies here equally well.

It is also worth emphasizing that the detection of a statistically significant correlation between two species' abundance patterns tells us nothing about the possible underlying reasons why this might be so. However, the detection of significant interspecific covariations can be extremely helpful in generating suitable hypotheses to explain such patterns, which then may lead to further experimental research.

12.2   PROCEDURES

Let the abundances of the ith species in N SUs be represented by the vector $y_i = [Y_{i1}, Y_{i2}, \ldots, Y_{iN}]$ and for the kth species by the vector $y_k = [Y_{k1}, Y_{k2}, \ldots, Y_{kN}]$. A positive correlation between these two species implies that when the abundance of one species increases in a SU that there is a corresponding increase in abundance in that SU by the other species. Similarly, a negative correlation implies that for an increase in one, there is a decrease in the other. If the vectors y_i and y_k have each been drawn from a normal distribution (the so-called *bivariate normal distribution*), then we may compute a *Pearson product-moment correlation coefficient* and test the hypothesis that a significant correlation does, in fact, exist. If the bivariate distributions are not normal, we compute a nonparametric coefficient, the *Spearman rank correlation*. The use of both coefficients presumes that a linear relationship exists between the species abundance vectors.

12.2.1   Pearson's Correlation

STEP 1.   STATE NULL HYPOTHESIS.   The null hypothesis is that species abundances are uncorrelated.

STEP 2.   COMPUTE PEARSON'S COEFFICIENT.   Pearson's correlation (r) between the ith and kth species is computed as

$$r(i, k) = \frac{\sum y_i y_k}{\sqrt{\sum y_i^2 \sum y_k^2}} \tag{12.1}$$

where

$$\sum y_i y_k = \sum_{j=1}^{N} Y_{ij} Y_{kj} - \left[\left(\sum_{j=1}^{N} Y_{ij} \right) \left(\sum_{j=1}^{N} Y_{kj} \right) \Big/ N \right]$$

$$\sum y_i^2 = \sum_{j=1}^{N} Y_{ij}^2 - \left[\left(\sum_{j=1}^{N} Y_{ij} \right)^2 \Big/ N \right]$$

$$\sum y_k^2 = \sum_{j=1}^{N} Y_{kj}^2 - \left[\left(\sum_{j=1}^{N} Y_{kj} \right)^2 \Big/ N \right]$$

given

Y_{ij} = the abundance of the ith species in the jth SU

Y_{kj} = the abundance of the kth species in the jth SU

The numerator of Eq. (12.1) is the covariance between species i and k, that is, a measure of how the abundances covary together. If this covariance is equal to 0, then $r = 0$, that is, the correlation between the two species is equal to 0. In other words, a change in the magnitude of the abundance of one species does not imply a reciprocal change in abundance of the other species. The abundances of the species may covary in either a positive or negative fashion, and the more they covary, the stronger the correlation. Pearson's correlation coefficient has a range from -1 (perfect negative correlation) to $+1$ (perfect positive correlation) and the coefficient is unitless.

STEP 3. TEST THE NULL HYPOTHESIS. The value of r as computed in Eq. (12.1) using sample data is an estimate of the population correlation coefficient, ρ. To test for the presence of a significant correlation between the two vectors, y_i and y_k, we test the null hypothesis that $\rho = 0$, versus the alternative that $\rho \neq 0$. In other words, if for example, $r = 0.47$ (as determined from a random sample), does a real correlation exist between y_i and y_k in the "population" or is the value of r simply a chance deviation from 0?

To test the null hypothesis, compare the absolute value of r with the critical value given in Rohlf and Sokal (1981, Table 25, p. 168) with df $= N - 2$ and a probability level of 5%. If $|r|$ exceeds this critical value, the null hypothesis is rejected.

Pearson's r has some properties that limit its usefulness as an affinity index (Boesch 1977): (1) it tends to exaggerate the overall importance of very large values in the data, (2) it may produce spurious correlations when there are many zeros in the data (see Section 12.5), (3) tests for significance of r assume an underlying normal frequency distribution, and (4) r assumes linear relationships between species abundances in the SUs.

12.2.2 Spearman's Rank Correlation

Often the bivariate population data in y_i and y_k are far from a normal distribution. If so, the procedures in the preceding section are inappropriate, and we turn to a nonparametric measure of covariation. One popular coefficient often used by ecologists is the Spearman rank correlation:

STEP 1. STATE NULL HYPOTHESIS. The null hypothesis is that the species "ranked" abundances are uncorrelated.

STEP 2. COMPUTE SPEARMAN'S COEFFICIENT. As the name suggests, the first task is to rank the abundance data in y_i and y_k in order from the largest to smallest values. This is readily illustrated with a simple example for species i and k and $N = 5$. If the abundance vectors are

$$y_i = [8, 6, 3, 0, 2]$$

$$y_k = [2, 14, 0, 6, 6]$$

then the ranked vectors are:

$$y_i(\text{ranked}) = [5, 4, 3, 1, 2]$$

$$y_k(\text{ranked}) = [2, 5, 1, 3.5, 3.5]$$

Note that the ranking of y_i is straightforward, whereas in y_k there is a "tie," since there are two abundances of 6. When ties occur, *each* member of a tie is given a rank that is the mean of the ranks that would have been given if no ties existed. Thus, in the above example, each 6 is assigned a rank of 3.5, which is the mean of 3 and 4, the ranks that would have been assigned if no ties were present.

The Spearman rank correlation coefficient, which we will represent as r_s, is then computed with Eq. (12.1) using $y_i(\text{ranked})$ and $y_k(\text{ranked})$. As with Pearson's r, r_s has no units and ranges from -1 to $+1$.

STEP 3. TEST NULL HYPOTHESIS. For the null hypothesis that the species ranked abundances are uncorrelated, we use the same test and rationale as used for Pearson's r. The same critical value tables can be used with relatively little error as long as $N > 10$ (Sokal and Rohlf 1981). Table 12.1 should be consulted if $N \leq 10$.

TABLE 12.1   *Significance levels of Spearman's rank correlation in small samples.*
The correlation is significant at the given probability level if the computed value of $|r_s|$
is greater than the table value for the sample size N (after Snedecor and Cohran 1973)

	Probability Level	
Sample Size (N)	$P = 0.05$	$P = 0.01$
≤ 4	—	—
5	1.00	—
6	0.886	1.00
7	0.750	0.893
8	0.714	0.857
9	0.683	0.833
10	0.648	0.794
≥ 11	a	a

[a] Use Table A 11, p. 557, Snedecor and Cochran (1973) or Table 25, p. 168, Rohlf and Sokal (1981).

12.2.3   Multiple Interspecific Covariations

As with interspecific associations in the previous chapter, we are often interested in how a number of species, in pairwise combinations, covary. Two ways of diagramming the covariation patterns among many species are:

DIAGRAM 1.   *COVARIATION COMPARISON MATRIX.*   As with interspecific associations, the results for interspecific covariations can be summarized in matrix form in a covariation comparison matrix as shown in Figure 12.1.

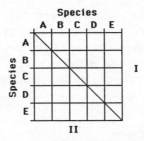

Figure 12.1   *Matrix of pairwise species covariations. The upper right triangle (I) could be used for correlations and the lower left triangle (II) could be used for covariance.*

DIAGRAM 2.   *COVARIATION PLEXUS DIAGRAM.*   The patterns of interspecific covariations can be artistically displayed in the form of a plexus diagram (see Section 11.2.3). The distance between species in a plexus diagram reflects their relative degree of positive covariation, that is, those that covary positively and significantly are displayed close together and those that negatively covary are displayed at a greater distance. For example, given pairwise interspecific covariations among five species, A, B, C, D, and E, Figure 12.2 illustrates how species B and D might group separately from species A, C, and E.

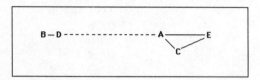

Figure 12.2   *Plexus diagram of species covariation patterns. Species B and D positively covary as do species A, C, and E (solid lines), but these groups are separated by negative covariations (dashed line).*

12.3   EXAMPLE: CALCULATIONS

Percent cover values for two species ($i = Festuca$ and $k = Cirsium$) were obtained from nine adjacent quadrats (SUs) on chalk grassland in England (Greig-Smith 1983):

$$y_i = [88.4, 84.8, 74.1, 73.2, 60.7, 57.1, 55.4, 43.8, 41.1]$$

$$y_k = [\ 8.9,\ \ 3.6, 19.6, 11.6, 31.3, 23.2, 18.8, 32.1, 33.0].$$

Based on presence–absence alone, we note that the species are completely associated (i.e., always occur together; see Chapter 11). Using these abundance data, we can compute whether or not a significant covariation in cover exists. On visual inspection of the two abundance vectors, it appears that there may be a tendency for the cover of *Cirsium* to increase as the cover of *Festuca* decreases.

12.3.1   Pearson's Correlation

The Pearson correlation coefficient is [Eq. (12.1)]

$$r(i, k) = \frac{\sum y_i y_k}{\sqrt{\sum y_i^2 \sum y_k^2}} = \frac{-1,285.06}{\sqrt{(2,278.81)(917.18)}} = -0.89$$

where

$$\sum y_i y_k = [(88.4)(8.9) + \cdots + (41.1)(33.0)]$$

$$- \{[(88.4 + \cdots + 41.1)(8.9 + \cdots + 33.0)]/9\}$$

$$= 10,421.95 - [(578.6)(182.1)/9] = -1,285.06$$

$$\sum y_i^2 = (88.4^2 + \cdots + 41.1^2) - [(88.4 + \cdots + 41.1)^2/9]$$

$$= 39,476.36 - [(578.6)^2/9] = 2,278.81$$

$$\sum y_k^2 = (8.9^2 + \cdots + 33.0^2) - [(8.9 + \cdots + 33.0)^2/9]$$

$$= 4,601.67 - [(182.1)^2/9] = 917.18$$

To test the significance of Pearson's correlation, we compare the computed value of r (0.89, ignoring sign) with the critical table value of 0.67 (df $= 7, P = 0.05$) obtained in Rohlf and Sokal (1981, Table 25, p. 168) or Snedecor and Cochran (1973, Table A11). Since $|r|$ exceeds the critical value, we reject the null hypothesis that species abundances are uncorrelated and conclude that there is a strong negative correlation in cover values for these two species.

12.3.2 Spearman's Correlation

The sample size in this example is fairly small (< 10) and, thus, the assumption of normality should be questioned. The nonparametric Spearman rank correlation is computed by first ranking vectors y_i and y_k in order from the highest to lowest cover values:

$$y_i(\text{ranked}) = [9, 8, 7, 6, 5, 4, 3, 2, 1]$$

$$y_k(\text{ranked}) = [2, 1, 5, 3, 7, 6, 4, 8, 9]$$

Now, from Eq. (12.1),

$$r_s(i, k) = \frac{\sum y_i(\text{ranked}) y_k(\text{ranked})}{\sqrt{\sum y_i^2(\text{ranked}) \sum y_k^2(\text{ranked})}} = \frac{-50}{\sqrt{(60)(60)}} = -0.83$$

where

$$\sum y_i(\text{ranked}) y_k(\text{ranked}) = [(9)(2) + \cdots + (1)(9)]$$

$$- \{[(9 + \cdots + 1)(2 + \cdots + 9)]/9\}$$

$$= 175 - [(45)(45)/9] = -50$$

$$\sum y_i^2(\text{ranked}) = (9^2 + \cdots + 1^2) - [(9 + \cdots + 1)^2/9]$$
$$= 285 - [(45)^2/9] = 60$$
$$\sum y_k^2(\text{ranked}) = (2^2 + \cdots + 9^2) - [(2 + \cdots + 9)^2/9]$$
$$= 285 - [(45)^2/9] = 60$$

Since $N < 10$, we compare the computed value of $r_s = -0.83$ to the critical value of 0.683 (Table 12.1; 7 df; $P = 0.05$) and conclude that r_s is significant (at $P = 0.05$, but not $P = 0.01$).

Note, that when the two vectors are identical in terms of the ranks assigned (as in this example), the denominator terms are identical and only one need be computed, that is,

$$\sum y_i^2(\text{ranked}) = \sum y_k^2(\text{ranked})$$

12.4 EXAMPLE: WISCONSIN FORESTS

Relative density classes for eight trees in ten upland forest sites in southern Wisconsin are given in Table 11.6a. To address the question of covariation in these values for each pairwise combination of species, the BASIC program SPCOVAR.BAS was used to compute the Pearson and Spearman correlation coefficients. The results are summarized in Table 12.2.

TABLE 12.2 *Pearson (upper-right triangle) and Spearman (lower-left triangle) correlation coefficients for tree species (SPP) abundance data (Table 11.6a) as computed by program SPCOVAR.BAS*

SPP	(1)	(2)	(3)	(4)	(5)	(6)	(7)	(8)
			Pearson Product-Moment Correlations					
(1)	1.00	0.67*	0.22	−0.02	0.30	−0.79**	−0.48	−0.84**
(2)	0.66	1.00	0.38	−0.57	0.02	−0.90**	−0.64*	−0.74*
(3)	0.16	0.26	1.00	0.13	0.12	−0.53	−0.71*	−0.62
(4)	−0.02	−0.56	0.20	1.00	0.01	0.35	0.02	0.05
(5)	0.32	−0.09	0.10	0.11	1.00	−0.30	0.06	−0.32
(6)	−0.77*	−0.86**	−0.45	0.40	−0.30	1.00	0.74*	0.90**
(7)	−0.48	−0.62	−0.71	0.01	0.07	0.63	1.00	0.77**
(8)	−0.85*	−0.73*	−0.58	0.01	−0.28	0.83*	0.74*	1.00
			Spearman Rank Correlations					

* = significant at 5% probability level.
** = significant at 1% probability level.

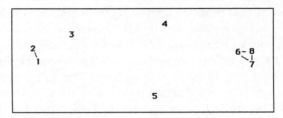

Figure 12.3 *Plexus diagram for eight tree species, upland forests, southern Wisconsin. Species connected by solid lines positively covary.*

A distinct pattern of species covariations is shown by the species covariation plexus diagram (Figure 12.3). The trees with low numbers (successional species) tend to covary negatively with trees with high numbers (forest climax species).

Of the 28 pairwise combinations, 10 correlations are significant using the Pearson correlation, but only 6 with the Spearman rank correlation. However, some cautionary notes are needed here. The sample size is small and, furthermore, it is highly unlikely that these density class data (Table 11.6a) are normally distributed. Given this, the nonparametric Spearman correlation is the appropriate covariation measure to use. Also, there are numerous zeros in this data set. In Section 12.5, we discuss options available in such cases: (1) eliminate the rare or infrequent species and/or (2) drop all double-zero matches. With regard to the latter, consider the case of species 7 (ironwood) and 8 (sugar maple). There are five double-zero matches involved in computing the r_s value of 0.74 (Table 12.2). If we eliminate these zeros from the calculations, the value of r_s for species 7 and 8 is -0.21. Thus, the student should be aware of the potential problems with zeros in a data set with small sample size.

Finally, since there are $S(S - 1)/2$ pairwise combinations of species, using a unidimensional significance test on each pair of species can lead to serious errors in interpretation (Legendre and Legendre 1983). The species are *not* independent and, thus, the tests are not independent. This problem is identical to the one we raised in Section 11.2.3 when using the chi-square to test for species associations. Multidimensional tests are available as discussed in the next section.

12.5 ADDITIONAL TOPICS IN INTERSPECIFIC COVARIATION

Ecological data sets are often characterized by the presence of many zeros. Recall that one of the conditions for selecting an association index in Chapter 11 was the double-zero matches be ignored (Condition 5, Section 11.2.2); otherwise, serious distortion may result. In computing the correlation coefficients, zeros are handled as "quantitative" data points. To avoid distortion, we have three options (Legendre and Legendre 1983): (1) eliminate the infrequent species, which should have minimal impact anyway on the search for meaningful affinities; (2) eliminate all zeros, that is, treat them as missing data; or (3) eliminate only double-zero matches. Depending on the nature of the data relative to the objectives, we recommend options 1 and 3. However, if the objective is to relate the covariation between species to patterns caused by competitive effects, only SUs containing both species should be used (Hurlbert 1969).

Instead of calculating species covariations using species pairs, we could pose the question as to whether the abundances of many S species in the community covary significantly among the N SUs. The variance ratio test described in Chapter 11 can be used to answer such a question. Schluter (1984) generalized the variance ratio test for presence–absence data to the case of abundance data. The variance ratio obtained can be used as an index of species covariation: a value significantly greater than 1 indicates that the species tend to covary positively and, conversely, a value significantly less than 1 indicates that the species covary negatively. Schluter provides a test statistic (W) that is approximately distributed as chi-square.

Other coefficients used as measures of species covariation include Kendall's tau (Greig-Smith 1983) and Morisita's (1959) correlation coefficient. The latter has been favorably compared to the commonly used Pearson's product-moment correlation (Hurlbert 1969). Morisita's correlation coefficient is relatively independent of average species abundance and diversity in the SUs (Wolda 1981).

In the previous section, we noted that there is a problem with individually testing the significance of a large number of correlations between pairwise combinations of species. Since the species are not independent, the significance tests are not independent, and a multidimensional, simultaneous test should be used (Legendre and Legendre 1983). Bartlett (1954) gives a statistic for testing the null hypothesis that the $S(S - 1)/2$ correlation coefficients are equal to zero, and Johnson and Wichern (1982) describe a multidimensional procedure for testing if the total number of correlation coefficients are significantly different from one another.

12.6 SUMMARY AND RECOMMENDATIONS

1. If abundance data for species are available (e.g., cover, biomass, numbers), we can assess the covariation in these data between species. This type of analysis is quite distinct from those techniques of species association presented in Chapter 11, which were based solely on species presence–absence data.

2. Correlation coefficients may be used to measure the degree of covariation in abundance data between two species. A positive correlation implies that for a given increase in abundance of one species, there is a corresponding increase in the other species. For negative correlation, an increase in one implies a decrease in the other. Of course, establishing the existence of a correlation does not imply causality.

3. If the data are drawn from a bivariate normal distribution, the Pearson correlation coefficient (Section 12.2.1) may be used. If the data are not normal, the nonparametric Spearman rank coefficient (Section 12.2.2) is recommended. Both coefficients assume that a linear relationship exists between the abundance of the two species being compared.

4. Many ecological community data sets are characterized by having numerous zeros, that is, many SUs with species absent (abundance = 0). We recommend that the student first eliminate infrequent species and then eliminate double-zero matches (SUs where abundances of both species are zero), when computing correlation coefficients. Failure to do this may result in meaningless (spurious) correlations (Section 12.2.3 and example in Section 12.4).

5. Caution should be exercised when interpreting the statistical significance of a large number of correlations between pairwise combinations of species since the species are not independent; thus, neither are the statistical tests (Section 12.5).

PART FIVE
COMMUNITY CLASSIFICATION

CHAPTER 13

Background

Given a set of objects and some measure of their resemblance to each other, we can define classification as the grouping or clustering of these objects based on their resemblance. Classification plays a fundamental role in many areas of science in the search for what Sokal (1974) terms the "natural" system. A natural system might be viewed as a reflection of those various processes that have led to the observed arrangement of the objects. For example, in ecology this "natural" system could be the end result of evolutionary processes.

The first step in the classification of ecological communities involves sampling. Artificial or natural sampling units (SUs) are used, and various types of data, both qualitative and quantitative, are obtained. These data may include lists of species present or some indication of their abundance (density, frequency, cover, biomass). Next, some measure of ecological resemblance between all pairs of SUs is computed in order to quantify their similarity or dissimilarity (what we term a Q-mode analysis, see Chapter 9). Finally, the SUs (objects) are grouped according to their resemblances; SUs in each group should have a number of common characteristics that set them apart from the SUs of other such groups. The objective is to demonstrate the relationships of the SUs to each other and, it is hoped, to simplify these relationships in order to be able to make general statements about the classes of objects that exist.

A study by Able and Noon (1976) is a good example of a potential classification problem. Their objective was to describe avian community structure along an elevational gradient in the Adirondack Mountains of New York. They found 44 species of birds distributed along a census belt transect that ranged from 400 to 1400 m elevation. Some species of birds were found in all

159

SUs along the gradient; others were found only within narrow limits of elevation. At several locations (ecotones) along the gradient, major changes in vegetation structure occurred, resulting in some natural upper and lower distribution limits for some of the bird species. A classification of the SUs in their study, based on species-abundance data, would reflect both the continuous and discontinuous distributions of all 44 bird species. The "natural" system of interest here, in a broad sense, is the grouping of the SUs that reflects the abundances (and amplitudes of abundances) of the major bird species along the elevational gradient.

Of course, homogeneous communities are not amenable to classification. Also, since some of the techniques presented in Chapters 15 and 16 will "classify" even a random data set, we have to be careful in our interpretation of these results. Simply because it may be possible to classify a data set, a spurious classification will not yield a meaningful ecological interpretation. Various philosophies of classification theory have been reviewed by Goodall (1970), Ratliff and Pieper (1982), Sokal (1974), and Whittaker (1978a, b); we highly recommend that students seriously interested in classification read these papers.

In the early days of ecology, classification of communities was largely intuitive, based on subjective decisions and qualitative descriptions. More recent trends have been toward objective methods of classification based on quantitative data. In the next few chapters, we will examine some of these objective methods. It is inevitable, however, that a degree of subjectivity remains in all classification studies; whereas a given classification method will yield unique results, an alternative method may yield different results and, consequently, subjective decisions must be made.

Some terminology used in the following chapters is briefly defined below:

1. Classifications may be either *hierarchical* or *reticulate*. As the name implies, in a hierarchical classification, groups at any lower level of a classification are exclusive subgroups of those groups at higher levels. In a reticulate classification (which we will not consider), groups are defined separately and, rather than hierarchically ordered, are linked together in a weblike network.

2. Classifications may be either *divisive* or *agglomerative*. In a divisive classification, the entire collection of SUs is divided and redivided, based on SU similarities, to arrive at the final groupings (i.e., picture an inverted tree). In an agglomerative classification, as its name implies, individual SUs are combined and recombined successively to form larger groups of SUs (the tree).

3. Classifications may be either *monothetic* or *polythetic*. In a monothetic classification the similarity of any two SUs or groups is based on the value of

a single variable, for example, the presence or absence of a single species. In a polythetic classification the similarity of any two SUs or groups is based on their overall similarity as measured by numerous variables, for example, species abundances.

In the subsequent treatment of classification methodologies, we make use of both qualitative data (e.g., presence–absence or two-state character data) and quantitative data (e.g., abundance or ordered multistate character data). In Chapter 14 indices of ecological resemblance pertaining to Q-mode analyses will be presented. In Chapter 15, the technique of normal association analysis is described; this is a monothetic, divisive classification model based on the presence–absence of species in SUs. In Chapter 16 a polythetic, agglomerative classification technique generally known as *cluster analysis* is described; this method is based on quantitative abundance data of species in SUs.

13.1   MATRIX VIEW

In studies of species affinity (Chapters 10, 11, and 12), the ecological data matrix was viewed across rows (Figure 9.2), that is, a R-mode analysis. In classification studies, the ecological data matrix may either be viewed across rows (Chapter 15) or down columns, that is, a Q-mode analysis (Figure 13.1). In either case, *the objective is the same: to classify SUs.*

Another way to view the relationship between R-mode (species) and Q-mode (SU) analysis is to conceptualize these relationships in a geometric way; this has lead to the term *hyperspace* (Williams and Dale 1965) when referring to R- and Q-mode studies. *Species hyperspace* is conceptualized as being S-dimensional, that is, one dimension for each species in the sample of S species. (Obviously, it is impossible for us to diagram the S-space beyond

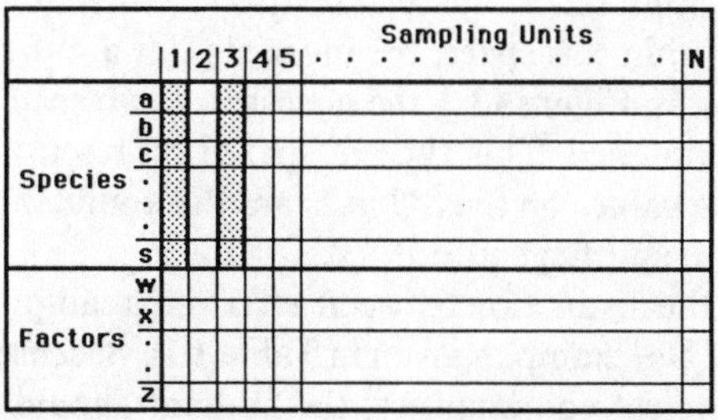

Figure 13.1   *The shaded area indicates the form of the ecological data matrix for measuring Q-mode resemblance. Interest is in pairwise SU similarities.*

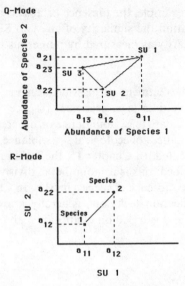

Figure 13.2 *Q- and R-mode analyses viewed geometrically. Q-mode is the representation of SUs in species space and R-mode is the representation of species in SU space. Note that $a_{i,j}$ is the abundance of the ith species in the jth SU. Adapted from Legendre and Legendre (1983).*

$S = 3$.) SUs are then positioned within this S-space based on the relative abundance of each species in a SU. The distance between the SUs in this S-space represents their similarity (or, alternatively, their dissimilarity; Chapter 14) to one another. An example is given in Figure 13.2, showing the location of three SUs (Q-mode) in the space of two species.

SU-hyperspace, on the other hand, is conceptualized as being N-dimensional, one dimension for each of the N SUs in the sample. The species are then positioned within this N-space in relation to their abundances; the closer two species are within this space, the more similar are their respective abundances in the SUs. In Figure 13.2, the position of two species (R-mode) in two SU-space is diagrammed. This type of spatial representation is much more artificial, since the values on the SU axes are the abundances of the species in the SUs (Legendre and Legendre 1983).

Finally, recall the distinction between a SU and a sample. A sample consists of a *collection* of SUs (examples given in Table 1.1). Because of the tremendous diversity of ecological communities, the student should be aware that the columns of the ecological data matrix may represent either individual SUs or samples. This is best illustrated by some examples from the literature.

In a study of benthic communities in Moreton Bay, Queensland, Australia,

Stephenson et al. (1970) conducted extensive dredging of the aquatic substrate. Their SU was a dredge that had a mouth 84 × 29 cm, a 2.5 m bag with 75 cm of mesh, and cutting edges inclined at 25°. These specifications are important because variations in any of these parameters would, most likely, affect the collection of bottom-dwelling animals. The speed of their boat was maintained at 0.6 km/hr and dredging was done for 2 minutes. The dredge catches were sorted to determine the macrobenthos species present. They collected 355 species in 400 dredge stations or locations throughout the bay. Their ecological data matrix of presence–absence data was 355 rows (species) by 400 columns (dredges or SUs), and both R-mode and Q-mode analyses were performed on these data.

In the next example, the columns of the data matrix represent samples rather than individual SUs. Huhta (1979) examined the changes in composition of soil arthropod communities in undisturbed and clear-cut forests north of Helsinki, Finland, from 1962 through 1965 and again in 1968. In each forest, samples were taken bimonthly. A sample consisted of four 25 × 25 cm quadrats of soil and litter, which were taken to the laboratory, and the total number of arthropod species and their relative abundances determined. These bimonthly data were combined into yearly samples for analysis. Thus, the data matrix in Huhta's study consisted of rows as soil arthropod species and columns representing the yearly samples (5 years); the entries in the matrix were number of individuals. Huhta proceeded to conduct a Q-mode analysis of these data to ascertain how the years differed in terms of the composition of the arthropod communities.

TABLE 13.1 Selected literature for examples of community classification studies using either association analysis (AA) or cluster analysis (CA)

Location	Community	Method	Reference
England	Chalk grassland	AA	Gittins 1965
NWT, Canada	Benthos	AA	Vilks et al. 1970
Virginia	Deciduous forest	AA	Madgwick and Desrochers 1972
Nigeria	Savanna	AA	Kershaw 1973
Australia	Forest–woodland	AA	Ashton 1976
Arizona	Desert grassland	AA	Fish 1976
North Sea	Marine benthos	CA	Stephenson et al. 1972
Australia	Rain forest	CA	Dale and Clifford 1976
Atlantic	Marine algae	CA	Lawson 1978
Puerto Rico	Rain forest	CA	Crow and Grigal 1979
New York	Deciduous forest	CA	Gauch and Stone 1979
England	Peat-bog	CA	Clymo 1980
NWT, Canada	Arctic tundra	CA	Thompson 1980
Australia	Rangelands	CA	Foran, et al. 1986

13.2   SELECTED LITERATURE

Classification studies in ecology are numerous. Two basic classification techniques are covered in the following chapters: (1) association analysis (AA, Chapter 15) and (2) cluster analysis (CA, Chapter 16). Clustering strategies, which are hierarchical, agglomerative, and polythetic, are by far the most commonly used; association analysis, a hierarchical, divisive, and monothetic method, is less commonly used. Some example of classification studies in the ecological literature are given in Table 13.1.

CHAPTER 14

Resemblance Functions

The ecologist is often faced with the task of making comparisons of plant and/or animal samples when addressing questions of community structure. These samples may be (1) obtained over various locations in the landscape, such as Able and Noon's (1976) study of bird distributions along an elevational gradient, or (2) obtained from the same location but at differing times, such as Livingston's (1976) comparisons of December and June fish catch data. In this chapter we describe some resemblance functions that quantify the similarity or dissimilarity between samples. The more similar samples are in species composition and quantity, the greater their resemblance, that is, the closer their ecological distance.

14.1 GENERAL APPROACH

Resemblance functions, as broadly defined by Sneath and Sokal (1973), quantify the similarity or dissimilarity between two objects based on observations over a set of descriptors. The objects of interest to the ecologist are SUs (sampling units or samples, see Chapter 13) and the descriptors are measures of species abundance (e.g., density, biomass). Thus, as defined, these resemblance functions involve a Q-mode analysis, that is, between SUs.

The distinction between Q-mode and R-mode analysis was made in Chapter 13 and the various resemblance functions used in these different modes was illustrated in Figure 9.1. In general, two types of Q-mode resemblance functions are distinguished: (1) similarity coefficients and (2) distance coefficients. Similarity coefficients vary from a minimum of 0 (when a pair of SUs are completely different) to 1 (when the SUs are identical). On the other hand, distance coefficients are the opposite; they assume a minimum value of 0 when a pair of SUs are identical and have some maximum value (in some cases infinity) when the pair of SUs are completely different. Hence, distance coefficients are also referred to as *dissimilarity coefficients*. In fact, a similarity index may always be represented as a distance, even if just by a simple transformation such as 1 − similarity (Legendre and Legendre 1983). Thus, distance may be thought of as the complement of similarity (Sneath and Sokal 1973).

Needless to say, the number of resemblance functions is large. In this chapter we limit our treatment to some of the more common similarity and distance measures used in Q-mode studies. However, this does not imply that some of the statistical and probability indices proposed to measure Q-mode resemblance (Legendre and Legendre 1983) might not be equally good, or, perhaps, even better with certain data sets (see Section 14.7). Distance coefficients are perhaps the most popular among community ecologists and, in our view, are the most straightforward in concept and application to community data.

14.2 PROCEDURES

14.2.1 Similarity Coefficients

Similarity coefficients are, by far, the most prolific indices in the ecological literature (Legendre and Legendre 1983). These indices are based solely on presence (indicated with a 1) or absence (indicated with a 0) data. Consider, for example, the presence–absence of 3 species in 3 SUs:

Species	SU (1)	(2)	(3)
A	1	1	0
B	1	1	0
C	1	0	1

In a Q-mode analysis, we are interested in the degree of similarity in species composition *between* each pair of SUs (columns of the data matrix). The more species two SUs share relative to their total species complements, the greater their ecological similarity. In this example, SU(2) contains two of the three species also found in SU(1), but has no species in common with SU(3).

Recall that in Section 11.2.2 we presented three indices (Ochiai, Dice, and Jaccard) based on presence–absence data that were used to measure the degree of association between species (an R-mode analysis, i.e., across the rows of the data matrix). These same three indices can also be used to compute a Q-mode similarity between SUs. The student should note that these are the *only types* of functions that we use to measure *both* Q-mode (sample similarity) and R-mode (species association) resemblance. The similarity between SU(1) and SU(2) in the above example, as measured by the Ochiai index [OI, Eq. (11.10)], Dice index [DI, Eq. (11.11)], and Jaccard index [JI, Eq. (11.12)] are:

$$OI_{1,2} = \frac{2}{\sqrt{2}\sqrt{3}} = 0.82$$

$$DI_{1,2} = \frac{4}{4 + 0 + 1} = 0.80$$

$$JI_{1,2} = \frac{2}{2 + 0 + 1} = 0.67$$

Since we covered the use of these indices in Chapter 11, they will not be presented again in this chapter.

14.2.2 Distance Coefficients

Three groups of distance measures are distinguished below: (1) E-group (the Euclidean distance coefficients), (2) BC-group (the Bray–Curtis dissimilarity index), and (3) RE-group (the relative Euclidean distance measures).

The following matrix notation is used in the equations presented below: X_{ij} represents the abundance of the ith species in the jth SU. For example, $X_{4,3}$ would be the abundance of the 4th species in the 3rd SU. As before, the community data matrix is composed of S species and N SUs.

14.2.2.1 E-Group Distances

DISTANCE 1. EUCLIDEAN DISTANCE (ED). This measure is the familiar equation for calculating the distance between two points SU_j and SU_k in Euclidean space:

$$ED_{jk} = \sqrt{\sum_{i=1}^{S} (X_{ij} - X_{ik})^2} \tag{14.1}$$

ED emphasizes the larger differences in abundances of species between SUs, since each species difference is squared and then summed. The final distance value is scaled down by taking the square root of the sum. The value of ED ranges from zero to infinity, as do all of the E-group measures.

DISTANCE 2. SQUARED EUCLIDEAN DISTANCE (SED). This measure is simply the square of ED:

$$SED_{jk} = \sum_{i=1}^{S} (X_{ij} - X_{ik})^2 \tag{14.2}$$

DISTANCE 3. MEAN EUCLIDEAN DISTANCE (MED). MED is similar to ED, but the final distance is on a smaller scale since the mean difference is used:

$$MED_{jk} = \sqrt{\frac{\sum_{i=1}^{S} (X_{ij} - X_{ik})^2}{S}} \tag{14.3}$$

DISTANCE 4. ABSOLUTE DISTANCE (AD). This measure is the sum of the absolute abundance differences taken over the S species:

$$AD_{jk} = \sum_{i=1}^{S} |X_{ij} - X_{ik}| \tag{14.4}$$

AD places less emphasis on larger differences than the previous three measures since differences in abundance are summed, but not squared. Thus, smaller differences are given relatively greater weight in the final distance. This distance measure is known as character difference in numerical taxonomy (Sneath and Sokal 1973).

DISTANCE 5. MEAN ABSOLUTE DISTANCE (MAD). The MAD is similar to AD but a mean distance is used rather than an absolute distance:

$$MAD_{jk} = \frac{\sum_{i=1}^{S} |X_{ij} - X_{ik}|}{S} \tag{14.5}$$

MAD is equivalent to mean character difference used in numerical taxonomy (Sneath and Sokal 1973).

14.2.2.2 BC-Group Distances. This group is represented by a single index first introduced into the ecological literature by Bray and Curtis (1957). This index remains very popular among ecologists. The first step is to compute the percent similarity (PS) between SUs j and k as

$$PS_{jk} = \left(\frac{2W}{A + B}\right)(100) \tag{14.6a}$$

where

$$W = \sum_{i=1}^{S} [\min(X_{ij}, X_{ik})]$$

$$A = \sum_{i=1}^{S} X_{ij} \quad \text{and} \quad B = \sum_{i=1}^{S} X_{ik}$$

Thus, PS between the jth SU and the kth SU is a numerator of twice the sum of the minimum (min) of the paired observations X_{ij} and X_{ik} (the "shared" species abundance between each pair of SUs) divided by a denominator of the total of all the species abundances for the two SUs. For any pair of SUs with identical species abundances, their similarity is complete, that is, PS = 100%.

The distance complement of PS is percent dissimilarity (PD), computed as

$$PD = 100 - PS \tag{14.6b}$$

PD may also be computed on a 0–1 scale as

$$PD = 1 - [2W/(A + B)] \tag{14.6c}$$

which is useful since it is more in line with the range of values assumed by many of the other distance indices. We will use PD as computed in Eq. (14.6c) in our calculations below.

14.2.2.3   RE-Group Distances.   This group contains distance indices that are expressed on standardized or relative scales.

DISTANCE 7.   RELATIVE EUCLIDEAN DISTANCE (RED).   This measure incorporates species abundance totals within each SU so that the final distance measure is standardized relative to differences in *total* SU abundances:

$$RED_{jk} = \sqrt{\sum_{i=1}^{S} \left[\left(\frac{X_{ij}}{\sum_{i}^{S} X_{ij}} \right) - \left(\frac{X_{ik}}{\sum_{i}^{S} X_{ik}} \right) \right]^2} \tag{14.7}$$

This equation is derived by applying Whittaker's (1952) relative transformation for absolute distance [Eq. (14.8)] to Euclidean distance as suggested by Orloci (1978). RED ranges from 0 to $\sqrt{2}$.

DISTANCE 8.   RELATIVE ABSOLUTE DISTANCE (RAD).   This measure applies Whittaker's (1952) relative abundance correction to AD (in the same sense that relative Euclidean distance "corrects" Euclidean distance):

$$\text{RAD}_{jk} = \sum_{i=1}^{S} \left| \left(\frac{X_{ij}}{\sum_i^S X_{ij}} \right) - \left(\frac{X_{ik}}{\sum_i^S X_{ik}} \right) \right| \tag{14.8}$$

RAD has a range from 0 to 2.

DISTANCE 9. CHORD DISTANCE (CRD). This measure puts greater importance on the relative proportions of species in SUs and correspondingly less importance on their absolute quantities. Technically, this is done by projecting the SUs onto a circle of unit radius through the use of direction cosines. The measure is then the *chord* distance between the two SUs after such a projection. We refer the student to Pielou (1984, p. 48) for a geometric illustration. Chord distance is given by

$$\text{CRD}_{jk} = \sqrt{2(1 - \text{ccos}_{jk})} \tag{14.9a}$$

where the chord cosine (ccos) is computed from

$$\text{ccos}_{jk} = \frac{\sum_{i=1}^{S} (X_{ij} X_{ik})}{\sqrt{\sum_i^S X_{ij}^2 \sum_i^S X_{ik}^2}} \tag{14.9b}$$

Note that, in the case of presence–absence data, this ccos is identical to Ochiai's coefficient. CRD, like RED, ranges from 0 to $\sqrt{2}$.

DISTANCE 10. GEODESIC DISTANCE (GDD). This measure is the distance along the *arc* of the unit circle (rather than the chord distance) after projection of the SUs onto a circle of unit radius:

$$\text{GDD}_{jk} = \arccos(\text{ccos}_{jk}) \tag{14.10}$$

GDD has a range from 0 to $\pi/2$ (i.e., 0 to 1.57).

To summarize the distances computed between all possible pairs of SUs based on any of the similarity or distance measures described previously, it is convenient to create a SU × SU matrix of distance (or similarity) values. Examination of this matrix quickly reveals the distance between any two SUs of interest. It is on this distance matrix that the clustering strategies of community classification operate (Chapter 16). We give an example of a distance matrix in Section 14.6.

14.3 EXAMPLE: CALCULATIONS

To illustrate the computations for the distance measures, the data in Table 14.1 will be utilized. From this simple data matrix of abundances for three

TABLE 14.1   *Community data matrix composed of three*
SUs with abundance data for three species (Spp)

Spp	SUs (1)	(2)	(3)
(1)	20	15	0
(2)	10	0	6
(3)	17	0	0

TABLE 14.2   *Differences (DIF), sums (SUM), and sums of squares (SSQ) needed*
for computing the distance between SUs 1 and 3

Spp	SU (1)	(3)	DIF $(1-3)$	DIF^2 $(1-3)^2$	SUM $(1+3)$
(1)	20	0	20	400	20
(2)	10	6	4	16	16
(3)	17	0	17	289	17
SUM =	47	6	41	705	
SSQ =	789	36			

species in three SUs, the computations for the distances between SUs 1 and 3 will be illustrated. For the computations, the differences, sums, and sums of squares within and between the three species in SUs 1 and 3 are needed (Table 14.2).

DISTANCE 1.   *EUCLIDEAN DISTANCE* [Eq. (14.1)]:

$$ED_{1,3} = \sqrt{[(20-0)^2 + (10-6)^2 + (17-0)^2]}$$
$$= \sqrt{(400 + 16 + 289)} = \sqrt{705} = 26.6$$

DISTANCE 2.   *SQUARED EUCLIDEAN DISTANCE* [Eq. (14.2)]:

$$SED_{1,3} = (400 + 16 + 289) = 705$$

DISTANCE 3.   *MEAN EUCLIDEAN DISTANCE* [Eq. (14.3)]:

$$MED_{1,3} = \sqrt{(705/3)} = \sqrt{235} = 15.3$$

When comparing these three related measures, note that the value of SED is much larger than that for ED and MED because the squared differences are summed. The species with the largest differences between SUs 1 and 3 (i.e., Spp. 1) receive the greatest weighting in the final distance value (e.g., 400 for Spp. 1 versus 289 for Spp. 3 and 16 for Spp. 2).

DISTANCE 4. ABSOLUTE DISTANCE [Eq. (14.4)]:

$$AD_{1,3} = |20 - 0| + |10 - 6| + |17 + 0| = 20 + 4 + 17 = 41$$

DISTANCE 5. MEAN ABSOLUTE DISTANCE [Eq. (14.5)]:

$$MAD_{1,3} = 41/3 = 13.7$$

When contrasting these two related measures with ED, SED, and MED, note that since differences are not squared, less relative importance is given to those species with the larger abundance differences (e.g., Spp. 1).

DISTANCE 6. BRAY–CURTIS DISSIMILARITY [Eq. (14.6c)]:

$$PD_{1,3} = 1 - [(2)(0 + 6 + 0)/(47 + 6)] = 1 - (12/53) = 0.77$$

DISTANCE 7. RELATIVE EUCLIDEAN DISTANCE [Eq. (14.7)]:

$$RED_{1,3} = \sqrt{[(20/47) - (0/6)]^2 + \cdots + [(17/47) - (0/6)]^2}$$
$$= \sqrt{(0.426 - 0)^2 + \cdots + (0.362 - 0)^2}$$
$$= \sqrt{0.181 + 0.619 + 0.131} = \sqrt{0.931} = 0.96$$

DISTANCE 8. RELATIVE ABSOLUTE DISTANCE [Eq. (14.9)]:

$$RAD_{1,3} = |(20/47) - (0/6)| + \cdots + |(17/47) - (0/6)|$$
$$= |0.426 - 0| + \cdots + |0.362 - 0|$$
$$= (0.426 + 0.787 + 0.362) = 1.57$$

DISTANCE 9. CHORD DISTANCE. First determine the cosine of the chord distance (ccos) using Eq. (14.9b):

$$ccos_{1,3} = [(20)(0) + (10)(6) + (17)(0)]/\sqrt{(789)(36)}$$
$$= (0 + 60 + 0)/\sqrt{28,404} = 60/168.5 = 0.356$$

Then the chord distance [Eq. (14.9a)]:

$$CRD_{1,3} = \sqrt{[2(1.0 - 0.356)]} = \sqrt{(2)(0.64)} = 1.13$$

DISTANCE 10. GEODESIC DISTANCE [Eq. (14.10)]:

$$GDD_{1,3} = \arccos[0.356] = 1.21$$

14.4 EVALUATION OF DISTANCE FUNCTIONS

In the previous section we presented 10 common distance functions. It is obvious that some of these are very similar to each other, while others seem to be quite different. In this section we have some examples of how these 10 functions perform on different data sets.

All possible (j, k) distances for the data set in Table 14.1 are shown in Table 14.3. Although SUs 2 and 3 share no species, the first five distance measures (the E-group) actually indicate that these two SUs are more similar (lower distance value) than either SUs 1 and 2 or SUs 1 and 3, which have one species in common. The final five distance measures (the BC- and RE-groups) do not give this unreasonable result. In fact, PD, RED, RAD, CRD, and GDD each give the same ranking of distances, that is, SUs 1 and 2 are the most similar and SUs 2 and 3 are the least similar, a more realistic result.

A first glance at Table 14.1 might intuitively suggest that the distance between SUs 1 and 2 would be larger than the distance between SUs 1 and 3.

TABLE 14.3 *Computed values for each of the 10 distance measures described in the text based on the data given in Table 14.1*

Distance Group	Distance Measure	Equation	SU(j, k)		
			(1, 2)	(1, 3)	(2, 3)
E	ED	14.1	20.4	26.6	16.2
	SED	14.2	414.	705.	261.
	MED	14.3	11.7	15.3	9.3
	AD	14.4	32.	41.	21.
	MAD	14.5	10.7	13.6	7.0
BC	PD	14.6	0.52	0.77	1.00
RE	RED	14.7	0.71	0.96	1.41
	RAD	14.8	1.15	1.57	2.00
	CRD	14.9	0.76	1.14	1.41
	GDD	14.10	0.78	1.21	1.57

For the species in common, the absolute difference in abundance between SUs 1 and 2 is 5 (Spp. 1), whereas this difference for SUs 1 and 3 is 4 (Spp. 2). However, the relative weights involved in computing these indices produces the result that the distance between SUs 1 and 2 is smaller than between SUs 1 and 3. Although the data set in Table 14.1 is artificial and simple, it does help to illustrate some of the difficulties when zero data are present in community data (the usual case) and the student should always proceed with caution.

RED and RAD express species abundances relative to the total abundance across all of the SUs. The effect of the RED and RAD expressions is to more equalize the importance of species relative to SUs with high and low total abundances. Two SUs with species in approximately the same proportions will tend to be more similar (i.e., have a close distance). Thus, if one is interested in measuring SU resemblance where species of high abundance in SUs with high total abundance will tend to be equally weighted with species of low abundance in SUs with low total abundance, the RED or RAD measures could be used.

Chord (CRD) and geodesic (GDD) distances compare species abundances relative to the abundance sums of squares for the SUs. Thus, as with RED and RAD, two SUs with species abundances in approximately the same proportions will be close in distance.

To illustrate further the performances of these resemblance functions, the data sets in Tables 11.4a and 14.4 were used to compute values for each of

TABLE 14.4   *Percentage abundance data for 11 species (A–K) in seven SUs.*
These data were used to compute rank correlations between the 10 distance indices
discussed in this section

Species	SUs						
	(1)	(2)	(3)	(4)	(5)	(6)	(7)
A	100	0	50	100	5	0	85
B	90	10	50	40	0	0	0
C	80	20	50	20	5	0	65
D	70	30	50	10	0	5	0
E	60	40	50	5	5	0	75
F	50	50	50	0	0	5	0
G	40	60	50	5	5	5	0
H	30	70	50	10	0	0	65
I	20	80	50	20	5	0	85
J	10	90	50	40	0	5	0
K	0	100	50	100	5	5	0

TABLE 14.5   Spearman rank correlations between the 10 distance measures.
Correlations above and below the diagonal are based on the data in Tables 11.4a and
14.4, respectively

Group		ED	SED	MED	AD	MAD	PD	RED	RAD	CRD	GDD
E	ED	1.0	0.99	0.99	0.99	0.99	0.27	−.25	−.24	−.22	−.22
	SED	1.0	1.0	0.99	0.99	0.99	0.27	−.25	−.24	−.22	−.22
	MED	1.0	1.0	1.0	0.99	0.99	0.27	−.25	−.24	−.22	−.22
	AD	0.86	0.86	0.86	1.0	1.0	0.27	−.26	−.24	−.21	−.21
	MAD	0.86	0.86	0.86	1.0	1.0	0.27	−.26	−.24	−.21	−.21
BC	PD	0.48	0.48	0.48	0.47	0.47	1.0	0.64	0.67	0.66	0.66
RE	RED	0.36	0.36	0.36	0.16	0.16	0.56	1.0	0.99	0.97	0.97
	RAD	0.31	0.31	0.31	0.20	0.20	0.63	0.92	1.0	0.97	0.96
	CRD	0.46	0.46	0.46	0.28	0.28	0.57	0.96	0.87	1.0	0.96
	GDD	0.46	0.46	0.46	0.28	0.28	0.57	0.96	0.87	1.0	1.0

these 10 distance functions over all pairwise combinations of SUs. Then, using
Eq. (12.1), Spearman rank correlation coefficients were computed between all
pairwise combinations of the 10 distance functions (Table 14.5). Distances
based on functions *within* the *E*-group (ED, SED, MED, AD, and MAD) and
the RE-group (RED, RAD, CRD, and GDD) are highly correlated; on the
other hand, the correlations *between* these groups are low. The Bray–Curtis
PD index had low correlations with the *E*-group, but fairly high correlations
(0.56–0.67) with the RE-group.

From these evaluations, we note the following:

1. In spite of the widespread popularity of the *E*-group distance measures,
   we do not recommend their use. It is clear from our results (Table
   14.3) that spurious results can occur. Wolda (1981) reached a similar
   conclusion.
2. Any of the RE-group functions appear to perform reasonably well. There
   seems little advantage in choosing any one over another (given their high
   correlation, Table 14.5), but we have found chord distance [Eq. (14.9)]
   to perform very satisfactorily over a diverse set of ecological data sets.
3. PD offers an alternative to the RE-group. This coefficient has been
   highly recommended by Beals (1984), based on his successful use of PD
   over a wide range of ecological studies.

14.5   EXAMPLE: PANAMANIAN COCKROACHES

This BASIC microcomputer program SUDIST.BAS (see accompanying disk)
was used to compute the distances between six Panamanian localities based

TABLE 14.6   *Distance measures between six Panamanian localities based on abundances for five cockroaches using SUDIST.BAS*

		Distance Measure									
Locality		E-Group					BC-Group	RE-Group			
(j)	(k)	ED	MED	SED	AD	MAD	PD	RED	RAD	CRD	GDD
1	2	52.2	23.3	2722	94	18.8	0.50	0.49	0.83	0.74	0.76
1	3	74.6	33.4	5571	115	23.0	0.97	0.74	1.28	1.10	1.16
1	4	75.0	33.5	5626	116	23.2	0.98	1.08	1.76	1.28	1.38
1	5	64.2	28.7	4127	101	20.2	0.76	0.25	0.47	0.34	0.34
1	6	54.5	24.4	2966	94	18.8	0.66	0.47	0.84	0.44	0.44
2	3	47.1	21.1	2219	69	13.8	0.95	0.60	0.89	0.87	0.90
2	4	47.2	21.1	2226	70	14.0	0.97	0.62	0.93	0.64	0.66
2	5	38.6	17.3	1489	55	11.0	0.63	0.40	0.57	0.56	0.57
2	6	38.5	17.2	1486	48	9.6	0.50	0.77	1.18	0.89	0.92
3	4	1.0	0.4	1	1	0.2	0.33	0.71	1.00	0.77	0.79
3	5	11.4	5.1	130	14	2.8	0.78	0.85	1.38	1.18	1.27
3	6	24.1	10.8	579	27	5.4	1.00	1.19	2.00	1.41	1.56
4	5	11.4	5.1	131	15	3.0	0.88	1.02	1.50	1.15	1.22
4	6	24.0	10.8	578	26	5.2	1.00	1.39	2.00	1.41	1.56
5	6	13.7	6.1	187	19	3.8	0.46	0.38	0.63	0.36	0.36

on the abundances of five cockroaches (data in Table 11.4a). The results are given in Table 14.6. Note the great differences in scale for the various distance measures. For example, squared Euclidean distance (SED) ranges up into the thousands, but also as low as one. Recall that SED and the other E-group measures (ED, MED, AD, and MAD) range from zero to infinity; this is because they increase as the number of species (S) increases. In contrast, recall that relative Euclidean distance (RED) and chord distance (CRD) have an upper limit of only $\sqrt{2} = 1.41$, relative absolute distance (RAD) has an upper limit of 2, and the geodesic distance (GDD) has an upper limit of $\pi/2 = 1.57$.

14.6   EXAMPLE: WISCONSIN FORESTS

Using the data matrix for eight trees in 10 upland forest sites, southern Wisconsin (Table 11.6a), the program SUDIST.BAS (see accompanying disk) was used to calculate all pairwise combinations of chord distances (CRD) between the 10 sites (Table 14.7). Recalling that the maximum value of CRD is 1.41 (for maximum dissimilarity), it is obvious that SUs 1 and 2 are the most similar, followed by SUs 9 and 10.

Often, these results are used for subsequent analyses, such as cluster analysis (Chapter 16). For such analyses, distances are conveniently used in the form

TABLE 14.7   *Chord distances between ten upland forest sites, southern Wisconsin in SU × SU matrix form (above diagonal)*

SUs	(2)	(3)	(4)	(5)	(6)	(7)	(8)	(9)	(10)
(1)	0.15	0.61	0.50	0.83	1.17	1.02	1.22	1.35	1.30
	(2)	0.64	0.50	0.85	1.16	1.02	1.22	1.33	1.29
		(3)	0.45	0.94	1.14	1.12	1.18	1.38	1.25
			(4)	0.57	0.95	0.90	1.05	1.28	1.18
				(5)	0.79	0.67	0.95	1.16	1.07
					(6)	0.74	0.41	0.80	0.80
						(7)	0.63	0.62	0.61
							(8)	0.52	0.53
								(9)	0.31

of a SU × SU comparison matrix (as illustrated by the 10 × 10 matrix of CRD distances in Table 14.7).

14.7   ADDITIONAL TOPICS ON RESEMBLANCE FUNCTIONS

Hubalek (1982) judged the "admissibility" of 43 similarity coefficients as Q-mode resemblance functions for presence–absence data based on five major "conditions." Hubalek suggested that only four "generally worked well" on a set of test data: (1) Jaccard's coefficient of community, (2) Dice's coincidence index, (3) Kulczynski's coefficient, and (4) Ochiai's coefficient. The Jaccard, Dice, and Ochiai coefficients were also highly recommended in an independent critical review of 20 similarity measures by Janson and Vegelius (1981).

Wolda (1981) examined the effects of sample size and species diversity on 22 measures of ecological resemblance, including product-moment and rank correlation coefficients and various measures based on information content. Wolda did not recommend either of the correlation coefficients as similarity indices. Of the information measures he examined, Wolda highly recommended Morisita's (1959) index because it proved to be independent of both sample size and diversity. When the data require prior log transformation, Wolda recommended Horn's (1966) simplified version of Morisita's index and Horn's index of overlap. However, Bloom (1981) found these two indices of Horn's to "diverge greatly from one another and from the theoretical standard" (the standard being based on a table of the area of a normal curve).

Ecologists have not made much use of the probability measures of resemblance, such as the indices described by Goodall (1964, 1966) and Feoli and Lagonegro (1983). This is due in part to the relatively complex and lengthy computations involved. The principal advantages of Goodall's index are: (1) it is on a scale from zero to one, (2) it is linear, and (3) it is applicable to both abundance and presence–absence data (Orloci 1978). The main disadvantage is that the probabilities for the similarity of any two SUs are based on all the SUs in the data set and, consequently, if SUs are either added to or taken from a data set, the probability-based similarity between the two given SUs will change.

Ecologists also have made little use of information indices, for example, Horn's (1966) index of overlap, for measuring the resemblance between SUs. Orloci (1978) describes some other information measures for SU resemblance. For excellent reviews of different resemblance functions, we refer the student to Boesch (1977), Campbell (1978), Clifford and Williams (1976), Goodall (1978b), Hubalek (1982), Orloci (1972, 1978), Pielou (1984), and Williams and Dale (1965).

Many ecological data sets are a mixture of quantitative information on the density, frequency, cover, biomass, and so forth, of species in SUs. Some species may be dominant in several SUs while absent in others. Also, there are species that may be only rarely found in the entire sample. To avoid the risk of overemphasizing the dominant species in the data analysis, ecologists often employ numerous standardizations or transformations of the data before computing ecological resemblances (Jensen 1978). A large number of types of transformations are possible, some of which we demonstrated above in Section 14.2.2.3. Chardy et al. (1976) used a logarithmic transformation before analyzing high-diversity plankton communities dominanted by only a few abundant species. Other transformations may also have been appropriate (e.g., angular and square-root transforms). Gauch (1982), Greig-Smith (1983), Jensen (1978), Noy-Meir (1973), Noy-Meir et al. (1975), and Orloci (1978) provide excellent overviews on the relative merits of ecological data transformations. Hajdu (1981), in a graphical comparison of 16 resemblance measures, found that standardization by SUs had undesirable effects on an ordered series.

Of course, the search for new (and the comparison of old) ecological resemblance functions will continue. For example, new similarity functions based on presence–absence (binary) data have been proposed and tested (Faith 1983, 1984), as well as a new way to compute distance (Bradfield and Kenkel 1987). The real ecological value will come from gained understanding of which measures are most robust when applied to classification and ordination procedures.

14.8   SUMMARY AND RECOMMENDATIONS

1. Two types of Q-mode (SU) resemblance functions may be distinguished: similarity coefficients and distance coefficients (Section 14.1).

2. We recommend the Ochiai, Dice, and Jaccard similarity coefficients for computing SU resemblance when the data consists of species presence–absence data (Section 14.2.1).

3. Three groups of distance functions based on abundance data may be distinguished: the E-group (Euclidean distance indices), the BC-group (represented by the Bray–Curtis dissimilarity index), and the RE-group (relative Euclidean distance indices) (Section 14.2.2).

4. We do not generally recommend the use of the E-group distance indices despite their widespread popularity. Although these indices have been of great heuristic value in ecology, there are various pitfalls in their use (Section 14.4).

5. For computing SU resemblances when the data consist of quantitative abundance data, we recommend chord distance from the RE-group of distance functions.

CHAPTER 15

Association Analysis

In this chapter we demonstrate the use of association analysis as a classification technique. Sampling units (SUs), e.g., quadrats and plots, are hierarchically sorted into homogeneous groups based on a divisive algorithm. A group is considered homogeneous when species associations (as determined from presence–absence data) disappear.

15.1 GENERAL APPROACH

Association analysis (AA) arose from the pioneering work of David Goodall (1953) and, perhaps because of the intuitive nature of the approach, was the first widely used classification scheme in plant ecology. The basic assumption underlying the technique of AA was stated by Goodall as: "If inspection of a set of quadrat data shows that the quantities of two species in the quadrats are [associated], it is probable that the distribution of one or both of these species over the area is non-uniform and that the habitat factors controlling them are also non-uniform." Thus, in AA it is implied that some species are more sensitive than others to environmental factors that control the structure of the community and therefore those species may form the basis of a classification (Coetzee and Werger 1975).

The method of AA as presented below is adapted from Williams and Lambert (1959). Given a collection of SUs, if there are species that exhibit significant associations with other species (based on computing a chi-square value for all possible 2×2 tables as described in Chapter 11), then the collection is considered to be "heterogeneous." Heterogeneity is reduced by subdividing the collection of SUs such that significant species associations disappear (or there is a reduced number of such associations). Therefore, by repeating this scheme over and over, the *within-group* heterogeneity of each group formed is reduced, while the *between-group* heterogeneity is increased. When species associations disappear by the formation of a group, we refer to that group as being "homogeneous." The end result is a classification of the SUs based on species associations. Association analysis has been successfully applied to a great variety of vegetation types, from forests to deserts (see Table 13.1).

15.2 PROCEDURES

Association analysis is a monothetic, divisive, hierarchical technique, as defined in Chapter 13. The aim of AA is to classify SUs and hence we considered AA to be a Q-mode analysis, although the procedure uses pairwise comparisons of species associations, which are based solely on species presence–absence data in SUs.

15.2.1 Computational Steps

STEP 1. FORM 2 × 2 CONTINGENCY TABLES.   For each of the possible pairwise combinations of species in a collection of SUs, a 2 × 2 contingency table is formed from the presence–absence data (see Chapter 11). For S species, there are $(S)(S-1)/2$ pairs of species.

STEP 2. COMPUTE THE CHI-SQUARE STATISTIC.   For each contingency table, a chi-square value is computed [Eq. (11.7)]. These values are compared to the critical table chi-square ($P = 0.05$, df = 1) of 3.84 for significance. If none of the $(S)(S-1)/2$ chi-square values is significant, then the collection (group) of SUs is considered homogeneous and the procedure is terminated. On the other hand, if any species have significant associations, the collection of SUs is considered heterogeneous and Step 3 is followed.

Numerous researchers have pointed out that the critical chi-square value of 3.84 is inappropriate for *simultaneous* chi-square tests and a larger value is needed (see Jensen et al. 1968). We discussed this problem in Section 11.2.3 and introduced Schluter's (1984) variance ratio test for simultaneously testing for significant associations. We return to this topic in Section 15.2.2.

STEP 3. SUM SIGNIFICANT CHI-SQUARES.   For each species, the sum of significant (> 3.84) chi-square values is computed. If a species has no significant associations, its sum would be zero. However, if a species has significant associations with, for example, four other species in the collection, the sum would be based on only those four values.

STEP 4. SELECT THE DIVISOR SPECIES. The species with the highest sum of significant chi-squares is considered the "divisor" species.

STEP 5. SORT SUs. The SUs are split into two groups based on the presence or absence of the divisor species. The result is a subdivision of the original group of SUs into two groups, one with the selected divisor species present and the other with the divisor species absent.

To test if the resultant groups are now themselves homogeneous, a new set of chi-squares must be constructed for *both* new groups (i.e., Step 1). If neither group of SUs has any significant species associations, then the objective has been achieved and each group can be considered to represent a homogeneous collection of SUs. However, if significant species associations remain in one or both groups of SUs, Steps 1–5 are followed until no significant associations are present, that is, this process is repeated until homogeneity is achieved in all groups. The product of AA is, therefore, the formation of groups of SUs where each group has greater homogeneity within than between other such groups.

A major issue in AA has been the question of whether or when "homogeneity" has been achieved. A range of stopping rules have been proposed and are discussed below.

15.2.2 Stopping Rules

Goodall's (1953) original AA method continued to subdivide groups until *no* significant chi-squares (> 3.84) remained. However, this approach often resulted in the formation of numerous small groups. To overcome this, Goodall pooled various combinations of the final groups and reexamined each of these, in turn, for homogeneity. Pooled groups that did not produce any significant species associations remained pooled as one final group of SUs. The main ramification of Goodall's pooling procedure is that it contradicts the hierarchical nature of association analysis.

Williams and Lambert (1959) chose to keep AA purely hierarchical by stopping further divisions at a level above that where excessive fragmentation occurs. Rather than stopping when no significant species associations were found, they reasoned that division should stop at a level proportional to the number of SUs (*N*) within the group being examined. They developed several mathematical criteria based on this principle.

Madgwick and Desrochers (1972) suggested that the stopping level should be proportional to the number of species (*S*), not the number of SUs (*N*). They argue that it is the number of pairwise species associations $[(S)(S - 1)/2]$ that leads to 3.84 being the inappropriate significance level (since 3.84 is the 5% significance level for a *single* pair test). They recommend using a significance level set by the number of *simultaneous* chi-square tests (species pairwise combinations) as tabulated by Jensen et al. (1968). As an example, for a simultaneous comparison of associations between 30 pairs of species the value of 9.885 is appropriate ($P = 5\%$).

We prefer to use a "trial-and-error" stopping rule. We use Goodall's criterion of subdividing groups until no significant chi-square values (> 3.84) remain, but, in the final analysis, subdivision is stopped at the level judged to produce the most ecologically meaningful groups of SUs. Using this rule, the chi-square value of 3.84 is treated as a convenient index, *not* as a statistical test of significance. Although this trial-and-error rule may not appear to be as objective as using the criteria set forth by others, it does enable us to avoid excessive fragmentation of groups. Also, we would not divide groups represented by only a few SUs. In a critical review of AA, Madgwick and Desrochers (1972) suggest 6 as the minimum number of SUs per group before subdivision is considered.

15.3   EXAMPLE: CALCULATIONS

Using the Panamanian cockroach data (6 locations, 5 species) from Table 11.4*b*, AA proceeds as follows:

STEPS 1 AND 2.   Results of tests for significant species associations are given in Table 15.1. The only chi-square value greater than 3.84 is between species

TABLE 15.1   *Chi-square values for testing the significance of species association between the five pairs of Panamanian cockroaches occurring in six localities. The computer program SPASSOC.BAS was used*

Species Pair	Association[a] Type	Chi-Square Value	Bias
1 2	−	6.00	*
1 3	−	2.40	*
1 4	−	0.24	*
1 5	+	0.60	*
2 3	+	2.40	*
2 4	+	0.24	*
2 5	−	0.60	*
3 4	+	0.60	*
3 5	+	0.38	*
4 5	+	0.60	*

[a] Sign indicates direction of the species association.

* The uncorrected chi-square values are biased because either the expected frequency in any cell in the 2 × 2 table is less than 1 and/or the expected frequency of more than two cells is less than 5.

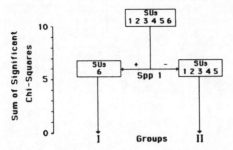

Figure 15.1 *Division of six Panamanian localities (SUs) using an association analysis based on the presence–absence of five cockroach species (Spp).*

1 and 2 ($\chi^2 = 6.0$). (Although this value is biased because of small sample size, we will continue the example.) Species 1 and 2 exhibit a negative association in that they do not occur together in any of the locations sampled (Table 11.4*b*).

STEP 3. Sum of significant chi-squares. From Table 15.1, species 1 and 2 have a sum of significant chi-squares equal to 6.0, and species 3, 4, and 5 have no significant associations.

STEP 4. Select the divisor species. Both species 1 and 2 tie for having the highest sum of significant chi-square values. In such cases, we arbitrarily select one of the two as the divisor. (With this example, the results are identical regardless of whether species 1 or 2 is selected.)

STEP 5. Divide the SUs on the basis of the presence or absence of the divisor species. The SUs sort into a group I with species 1 present (SU 6) and into a group II with species 1 absent (SUs 1–5) (Figure 15.1). Each group is now independently examined for homogeneity by recalculating pairwise species associations. Since species 1 is now absent in all SUs of group II, it no longer enters into these new calculations. No further divisions are attempted because group II has only five SUs.

15.4   EXAMPLE: WISCONSIN FORESTS

An association analysis of the Wisconsin forest data (8 species, 10 sites) was performed using the BASIC program NASSOC.BAS (accompanying disk), which follows the steps for an AA as outlined in Section 15.2.

In Section 11.5 we accepted the hypothesis of no association among the eight tree species. This was based on the variance ratio test for simultaneous pairwise associations. These results suggest that an AA is not warranted with these data, since no *overall* species association exists. However, the computed

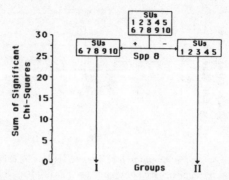

Figure 15.2   *Division of 10 upland forest sites (SUs), southern Wisconsin, using an association analysis based on the presence–absence of eight tree species (Spp).*

chi-square values that were given in Table 11.7 indicate a number of significant associations (all biased because of small sample size) between *pairs* of species. Following our chosen criteria for "homogeneity," we proceed with the AA.

The results for an AA are illustrated in Figure 15.2 and reveal the collection of 10 SUs is not homogeneous. Species 8 had the largest sum of significant chi-squares (26.7—the student can quickly verify this result from Table 11.7). Division on the presence or absence of species 8 produces two groups with five SUs each (Group I contains SUs 6, 7, 8, 9, and 10 and Group II contains SUs 1, 2, 3, 4, and 5). Groups I and II are not further subdivided, since the number of SUs in each group is less than six.

If we now reexamine the Wisconsin forest data, we can see that these results make some ecological sense. The five SUs representing Group I (SUs 6–10) are characterized by climax forest species, that is, sugar maple (the "divisor" species), basswood and ironwood, while the SUs in Group II (SUs 1–5) tend to be successional forest species, that is, the oaks (Peet and Loucks 1977). Although we have cautioned the student throughout this book against the use of small data sets, this example serves to illustrate the appeal of AA as a simple classification model.

15.5   ADDITIONAL TOPICS IN ASSOCIATION ANALYSIS

In this chapter we focused on a type of AA called *normal association analysis* because it leads to the classification of SUs based on the presence–absence of species. By simply turning the table (i.e., transposing the data matrix) and conducting the same AA, we can form groupings of species. In the ecological literature this type of AA is referred to as *inverse association analysis* (Williams and Lambert 1961). Species with similarities in habitat (SU) occurrences will tend to group together. Furthermore, the results of normal and inverse association analysis can be cross-tabulated to search for *nodes* of coincidence between SU and species groups (nodal analysis—Boesch 1977, Lambert and Williams 1962, and Noy-Meir 1971).

Madgwick and Desrochers (1972) recommend that rare species be excluded from AA. Their study indicated that the inclusion of rare species increased the chance that the frequency within any one cell of a 2 × 2 contingency table will be less than 3, which results in undue splitting of groups. They also suggest the use of either corrected chi-square (although they found that Yates' correction "overcorrects") or Fisher's exact test, which was also suggested by Pielou (1977). The option of using Yates' correction factor is available in our program NASSOC.BAS, and we encourage the student to examine the effect of choosing alternative values for chi-square on the results of AA.

Using 39 species and 167 SUs (forest plots), Madgwick and Desrochers (1972) demonstrated a problem with all monothetic classification techniques: the misclassification of SUs due to chance occurrences of the divisor species. Since the technique of AA is based on presence–absence data, the chance occurrence of a divisor species in an SU (which might, in fact, be dominated by other species) can lead to spurious and misleading results. By testing their AA classification with discriminant analysis (see Chapter 23), they found that 27 of the 167 plots were misclassified due to positive selection, that is, on the basis of the divisor species being present, while only five plots were misclassified due to the divisor species being absent.

One of the big advantages of the monothetic, divisive classification strategy is its computational efficiency when large numbers of SUs are involved, as in large ecological surveys. Starting with all SUs in one group and then subdividing until satisfactory homogeneous groups of SUs are obtained is computationally more efficient than starting with all SUs separate and then combining and recombining (the agglomerative strategy, Chapter 16). There have been other monothetic and divisive techniques described, including divisive information analysis (Lance and Williams 1968) and group analysis (Crawford and Wishart 1967, 1968), but the simple AA procedure described above remains as the basic method. Polythetic, divisive techniques have also been proposed (Greig-Smith 1983), which would reduce the misclassification problem of the monothetic method.

15.6 SUMMARY AND RECOMMENDATIONS

1. Association analysis (AA) is a classification scheme that is based on reducing "heterogeneity" within a group of SUs. A group is considered to be

heterogeneous if significant species associations exist (based on all pairwise species 2 × 2 tables, Chapter 11). Heterogeneity is reduced by subdividing the collection of SUs so that significant species associations disappear.

2. Since all possible pairwise species 2 × 2 tables are examined for the presence of significant associations, we recommend the use of the variance ratio test (Chapter 11) before conducting an AA. If this simultaneous test suggests that no significant species associations exist in the data, an AA would be of questionable use. The student also should always be cognizant of the fact that methods such as AA will produce a classification, but it may be fortuitous.

3. As a stopping rule in AA, we recommend that a "trial-and-error" approach be used in conjunction with Goodall's criterion for subdividing SUs into groups until no chi-square greater than 3.84 ($P = 0.05$, df $= 1$) exists, that is, subdivisions should stop at the point where the most ecologically meaningful classification is obtained. No statistical significance is implied here.

4. We also recommend six as the minimum number of SUs in a group, that is, groups with less than six SUs should not be divided.

CHAPTER 16

Cluster Analysis

Cluster analysis (CA) is a classification technique for placing similar entities or objects into groups or "clusters." The cluster analysis models we present in this chapter are used to place similar samples into clusters, which are arranged in a hierarchical treelike structure called a *dendrogram*. These clusters or groups of SUs may delimit or represent different biotic communities.

16.1 GENERAL APPROACH

Given a set of objects and some measure of their resemblance to each other, we defined classification (Chapter 13) to be the "sorting" of these objects into groups or clusters. Cluster analysis is a technique that accomplishes this sorting. The objects of concern here are ecological samples or SUs (e.g., plots, transects, quadrats). CA is actually a general term that refers to a large number of algorithms that differ mainly in their treatment of cluster formation.

We first present the general approach to CA before detailing procedures. Initially, we must compute a Q-mode resemblance between the SUs. Although numerous resemblance functions could be used, we restrict our coverage to distance measures (see Chapter 14), because of their heuristic value in CA (Sneath and Sokal 1973). The distances between all pairwise combinations of SUs in a collection are summarized into a SU $\times$ SU distance or D matrix (e.g., Table 14.7) and the various CA strategies operate on this D matrix.

The CA models we describe in this chapter are agglomerative (Chapter 13): they begin with a collection of N individual SUs and progressively build groups or clusters of similar SUs. During each clustering cycle, only *one pair*

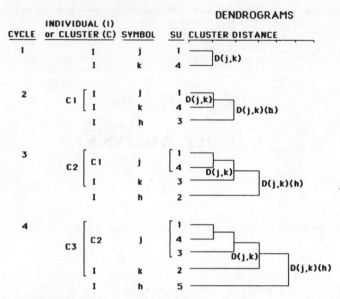

Figure 16.1   *Illustration of the Lance and Williams (1967) combinatorial clustering method on five SUs [see Eq. (16.1)].*

of entities may be joined to form a new cluster. This pair may be (1) an individual SU with another individual SU, (2) an individual with an existing cluster of SUs, or (3) a cluster with a cluster. Hence, the term *pair-group CA* is applied.

A general example of the pair-group approach is illustrated in Figure 16.1. For this example we use five SUs and, hence, 10 pairwise $[N(N-1)/2]$ distance values form the D matrix. The first step in all pair-group CA strategies involves searching the D matrix for the smallest distance value between two individual SUs. In Figure 16.1, this is shown to be between SUs 1 and 4, represented by the symbols j and k, respectively. Hence, the first cluster is formed at a distance $D(j,k)$ and this can be diagrammed using a dendrogram (Figure 16.1, Cycle 1). The initial collection of five SUs is now reduced to one cluster ($C_1 =$ SUs 1 and 4 joined) and three individual SUs (2, 3, and 5). The distance between this cluster and each of these three remaining SUs must now be computed. Special equations have been developed for this type of computation and a general one by Lance and Williams (1967), called the *linear combinatorial equation*, is given below.

The linear combinatorial equation takes the form

$$D(j,k)(h) = \alpha_1 D(j,h) + \alpha_2 D(k,h) + \beta D(j,k) \qquad (16.1)$$

TABLE 16.1 Parameter values for α_1, α_2, and β in the Lance and Williams' combinatorial equation [Eq. (16.1)] for different hierarchical clustering strategies. Names of the strategies follow Sneath and Sokal (1973)/Lance and Williams (1967). The number of SUs in the jth and kth groups are $t(j)$ and $t(k)$, respectively, and the number of SUs in the combined group (j, k) is $t(j, k)$

Strategy	α_1	α_2	β
Centroid (unweighted)/ centroid	$t(j)/t(j,k)$	$t(k)/t(j,k)$	$-t(j)t(k)/t(j,k)$
Centroid (weighted)/ median	$\frac{1}{2}$	$\frac{1}{2}$	$-\frac{1}{4}$
Group mean/unweighted pair-grouping method	$t(j)/t(j,k)$	$t(k)/t(j,k)$	0
Flexible	0.625	0.625	-0.25^a

$^a\beta$ is flexible with this strategy under the constraint that $\alpha_1 + \alpha_2 + \beta = 1$ and that $\alpha_1 = \alpha_2$.

where the distance between the new cluster (j, k) formed from the jth and kth SUs and a third hth SU or group of SUs can be calculated from the known distances $D(j, k)$, $D(j, h)$, and $D(k, h)$ and the parameters α_1, α_2, and β. For example, the distance between SU 3 and the cluster represented by SUs 1 and 4 (Figure 16.1, Cycle 2) is given by

$$D(1,4)(3) = \alpha_1 D(1,3) + \alpha_2 D(4,3) + \beta D(1,4) \qquad (16.2)$$

The different clustering strategies differ only in their values for α_1, α_2, and β (Table 16.1), which are the weights for determining the new distances (more on this below).

The relationships between the SUs and the formation of new clusters, as given in Eq. (16.1), are depicted in Figure 16.1 [the j's, k's and h's are those in Eq. (16.1)]. From Figure 16.1:

1. Given N SUs in a collection, there are $N - 1$ cycles in CA. In this example, there are four cycles.
2. In cycle 1, two individual SUs (represented by I's) are joined to form a cluster. The distance at which SU 1 (symbol j) and SU 4 (symbol k) form a cluster is given by $D(j, k)$, the value from the D matrix.
3. In cycle 2, SU 3 (symbol h) joins the cluster formed in cycle 1 (symbol C_1). The j and k are SUs 1 and 4, respectively, and the cluster distance between SU 3 and C_1 is $D(j, k)(h)$.
4. In cycle 3, SU 2 (symbol h) joins the cluster formed in cycle 2 (symbol C_2). Note that j now represents cluster C_1, and k is the latest SU to join C_1.

5. In cycle 4, SU 5 (symbol h) joins cluster C_3. In Eq. (16.1), j is cluster C_2 (SUs 1, 3, 4) and k is SU 2.

As previously mentioned, the α's and β in Eq. (16.1) determine the "weighting" of the distances. Depending on the weighting scheme used, the resultant cluster formation will vary. In some cases, the differences are dramatic. Four specific weighting schemes, the centroid (unweighted and weighted), group-average, and flexible, are given in Table 16.1.

The concept of weighting is probably best illustrated by an example. Returning to the example in Figure 16.1, for group-mean weighting the cluster distance between cluster C_3 (where j = SUs 1, 4, and 3 and k = SU 2) and SU 5 is given by [Eq. (16.1)]

$$D(1, 4, 3; 2)(5) = \tfrac{3}{4}D(1, 4, 3; 5) + \tfrac{1}{4}D(2, 5) \qquad (16.3)$$

where $\alpha_1 = \tfrac{3}{4}$, $\alpha_2 = \tfrac{1}{4}$, and $\beta = 0$ (from Table 16.1).

The group mean clustering strategy (the unweighted pair-group method with arithmetic averages), effectively computes the mean of all distances between SUs of one group to the SUs of another and, hence, is unweighted. On the other hand, for the weighted centroid strategy (Table 16.1), the combinatorial equation is

$$D(1, 4, 3; 2)(5) = \tfrac{1}{2}D(1, 4, 3; 5) + \tfrac{1}{2}D(2, 5) - \tfrac{1}{4}D(1, 4, 3; 2) \qquad (16.4)$$

which weights all fused groups as coequal regardless of differences in the number of SUs in each group. Also, note that in the centroid strategy, once a group is formed, it is replaced by its mean and intercluster distances are those distances between these means or centroids.

16.2 PROCEDURES

The various clustering procedures operate on the D matrix of all possible pairwise combinations of distances between SUs (e.g., Table 14.7). Any of the distance measures presented in Chapter 14 could be used. It is assumed that there are a total of N SUs in the collection.

STEP 1. OBTAINING THE INITIAL GROUP. The $N \times N$ D matrix is searched for the smallest distance value between a pair of SUs. This pair represents the two most similar SUs in the collection. These two SUs are joined (e.g., Figure 16.1).

STEP 2. REDUCTION OF THE D MATRIX. There are now $N - 1$ entities in the collection, in other words, one group composed of two SUs and the remaining $N - 2$ individual SUs. The distances between the new group and these remaining SUs is computed from Equation 16.1. A new reduced D' matrix is formed which is now $(N - 1) \times (N - 1)$.

STEP 3. SEARCH THE REDUCED D' MATRIX. Just as in Step 1, the new D' matrix is searched for the lowest distance value in order to identify the next new group to form.

STEP 4. REPEAT STEPS 2 AND 3 UNTIL ALL SUs ARE JOINED INTO ONE GROUP. This will take a total of $N - 1$ cycles since only a single pair (SU–SU, cluster–SU, or cluster–cluster) may be clustered during any one computational cycle. The number of entities present at the beginning of any cycle (Step 2) is $N - C$, where C is the cycle number.

The final problem in CA is one of identifying specific groups or communities once the clustering is completed. The dendrogram, as shown in Figure 16.1, can be examined for groupings of SUs. While this is largely a subjective decision, there have been some recent attempts to render this decision somewhat more objective (e.g., Hill 1980, Popma et al. 1983, Ratliff and Pieper 1981, Rohlf 1974, 1982). These objective procedures will be discussed in Section 16.6. A general guideline is that one does not divide so finely that one ends up with a large number of fragmentary and uninterpretable groups.

16.3   EXAMPLE: CALCULATIONS

The Lance and Williams (1967) combinatorial linear model [Eq. (16.1)] is illustrated by its application to a D matrix of Euclidean distances (Table 16.2). These distances were computed from the contrived data for abundances of three species in five SUs (Table 11.3a). The reconstructions of D after each clustering fusion are also given in Table 16.2. The flexible CA strategy is illustrated.

STEP 1. OBTAIN THE INITIAL GROUP. The smallest Euclidean distance in the D matrix is 1.41 between SUs 2 and 3. Hence, these two SUs are the first group formed and this can be depicted in a dendrogram as shown in Figure 16.2, clustering Cycle 1.

STEP 2. REDUCTION OF THE D MATRIX. The distance between this new group (2, 3) and the three remaining SUs is computed using Eq. (16.1) as

$$D(2, 3)(1) = (0.625)(4.69) + (0.625)(5.10) - (0.25)(1.41)$$
$$= 2.93 + 3.19 - 0.35 = 5.77$$

TABLE 16.2 *The D matrix of Euclidean distances between five SUs based on the data in Table 11.3a. Only the upper-right triangle is shown: (a) original D matrix, and (b)-(d) reduced D matrices after successive SU fusions*

		Sampling Unit (SU)			
(a)		(2)	(3)	(4)	(5)
	(1)	4.69	5.10	3.00	2.24
$D =$ (2)			1.41	2.24	5.74
(3)				3.00	5.92
(4)					3.74
(b)		(2, 3)	(4)	(5)	
	(1)	5.77	3.00	2.24	
$D' =$ (2, 3)			2.93	6.94	
(4)				3.74	
(c)		(2, 3)	(4)		
$D'' =$ (1, 5)		7.39	3.66		
(2, 3)			2.93		
(d)		(2, 3, 4)			
$D''' =$ (1, 5)		6.18			

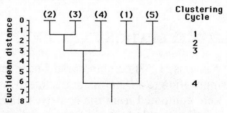

Figure 16.2 *Dendrogram of the clustering of five sampling units using Euclidean distance and the flexible strategy ($\beta = -0.25$).*

$$D(2, 3)(4) = (0.625)(2.24) + (0.625)(3.00) - (0.25)(1.41)$$

$$= 1.40 + 1.88 - 0.35 = 2.93$$

$$D(2, 3)(5) = (0.625)(5.74) + (0.625)(5.92) - (0.25)(1.41)$$

$$= 3.59 + 3.70 - 0.35 = 6.94$$

The reduced D' matrix is shown in Table 16.2b. Note that the distances between the unclustered SUs remain unchanged.

STEP 3. SEARCH THE REDUCED D' MATRIX. The smallest distance in D' is 2.24 between SUs 1 and 5. Hence, these two SUs are the next cluster formed as shown in Figure 16.2, clustering cycle 2.

STEP 4. REDUCTION OF THE D' MATRIX. The distance between this new cluster and the remaining SU (4) and group (2, 3) is computed as

$$D''(1,5)(2,3) = (0.625)(5.77) + (0.625)(6.94) - (0.25)(2.24)$$

$$= 3.61 + 4.34 - 0.56 = 7.39$$

$$D''(1,5)(4) = (0.625)(3.00) + (0.625)(3.74) - (0.25)(2.24)$$

$$= 1.88 + 2.34 - 0.56 = 3.66$$

Note that the reduced D' matrix from the previous cycle is used to obtain all new distances. This next reduced D'' matrix is shown in Table 16.2c.

STEP 5. SEARCH THIS REDUCED D'' MATRIX. The smallest distance value in D'' is 2.93 between the group represented by SUs 2 and 3 and SU 4. Hence, these three SUs are joined to form a new cluster at a distance of 2.93 as shown in Figure 16.2, clustering cycle 3.

STEP 6. REDUCTION OF THIS D'' MATRIX. The distance between this new cluster of three SUs and the only remaining entity, a group composed of SUs 1 and 5, is computed as

$$D'''(2,3;4)(1,5) = (0.625)(7.39) + (0.625)(3.66) - (0.25)(2.93)$$

$$= 4.62 + 2.29 - 0.73 = 6.18$$

STEP 7. The final reduced matrix D''' is shown in Table 16.2d and the final fusion joins all of the SUs together at a Euclidean distance of 6.18. This is illustrated in Figure 16.2, clustering cycle 4.

16.4 EXAMPLE: PANAMANIAN COCKROACHES

The BASIC program CLUSTER.BAS (see accompanying disk) was used to compute a CA on the Panamanian cockroach data set of Table 11.4a. There are five species of cockroaches in six locations, and relative Euclidean distance (RED, see Section 14.2.2.3) is the resemblance function used along with the flexible clustering strategy (see Table 16.1). A summary of the output from the BASIC program is given in Table 16.3. Note that a cluster is referenced by the SU with the lowest numerical value (for example, at cycle 2, the cluster consisting of SUs, 1, 5, and 6, is referred to as "cluster 1").

TABLE 16.3 Program CLUSTER.BAS results giving (a) distances between the six Panamanian localities (SUs), and (b) clustering of the localities

(a) Relative Euclidean distances (D matrix)

SUs	(2)	(3)	(4)	(5)	(6)
(1)	0.49	0.74	1.08	0.25	0.47
(2)		0.60	0.62	0.40	0.77
(3)			0.71	0.85	1.19
(4)				1.02	1.39
(5)					0.38

$D = $

(b) Clustering by the flexible strategy with $\beta = -0.25$

Clustering Cycle	No. of Groups	Clustering Level	Reference SU[a]	SUs in the Group
1	5	0.25	1	5
2	4	0.47	1	5, 6
3	3	0.60	2	3
4	2	0.68	2	3, 4
5	1	1.43	1	All SUs form one group

[a] The lowest numerical value of SUs in group.

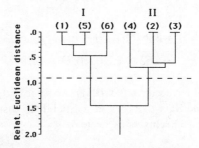

Figure 16.3 Dendrogram for the cluster analysis of six Panamanian localities using RED and the flexible strategy. The horizontal dashed line shows the arbitrary division line for defining clusters I and II.

The pattern of clustering for the six locations (SUs) is summarized in the dendrogram in Figure 16.3. To illustrate how these results might be interpreted, we have arbitrarily used a "cutoff" distance of 0.9 (shown as a horizontal dashed line in Figure 16.3). At this level of resemblance there are two distinct clusters: I (SUs 1, 5, and 6) and II (SUs 2, 3, and 4). Referring to Table 11.4a, it can be seen that SUs 1, 5, and 6 are largely dominated by a single species, *Latindia dohrniana*. Of course, this is a simplified data set and used only for illustration (see Section 16.6 for a further discussion of interpreting CA results).

16.5   EXAMPLE: WISCONSIN FORESTS

The Wisconsin forest community data in Table 11.6a was used with the BASIC program CLUSTER.BAS. The results of CA on these 10 upland forest sites (with eight trees) using each of the four strategies are given in Table 16.4. Chord distance [CRD, Eq. (14.9)] was used.

The results for the flexible strategy are summarized in a dendrogram (Figure 16.4). The two arbitrary dashed lines, at chord distances of 1.0 and 1.5, can be used as reference points for identifying clusters. At a distance of 1.0, three clusters emerge: *I* (SUs 1, 2, 3, and 4), *II* (SUs 5 and 7), and *III* (SUs 6, 8, 9, and 10). At the higher chord distance of 1.5, clusters *II* and *II* fuse, forming a single cluster. Thus, the four sites dominated by bur and black oak (SUs 1–4) form a cluster distinct from the remaining six sites (SUs 5–10), which are characterized by basswood and sugar maple (see Table 11.6a).

From a comparison of each CA strategy (Table 16.4), it can be seen that the results for the centroid methods and the group average are essentially identical to those previously described for the flexible strategy. The major differences are in the clustering of SUs 5 and 7. These SUs are, in fact, somewhat intermediate between Clusters *I* and *III* (Figure 16.4), that is, they have species characteristic of both clusters (see Table 11.6a).

These results illustrate an important point. Our experience suggests that

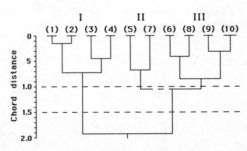

Figure 16.4   *Dendrogram for cluster analysis of 10 upland forest sites, southern Wisconsin, using CRD and the flexible strategy. The horizontal dashed lines represent reference points for delimiting clusters I, II, and III.*

TABLE 16.4 Program CLUSTER.BAS results giving (a) chord distances between the 10 Wisconsin forest sites (SUs), and clustering of 10 sites by the (b) weighted centroid, (c) unweighted centroid, (d) group-average, and (e) flexible strategies

(a) Chord distances (D matrix)

SUs	(2)	(3)	(4)	(5)	(6)	(7)	(8)	(9)	(10)
(1)	0.15	0.61	0.50	0.83	1.17	1.02	1.22	1.35	1.30
(2)		0.64	0.50	0.85	1.16	1.02	1.22	1.33	1.29
(3)			0.45	0.94	1.14	1.12	1.18	1.38	1.25
(4)				0.57	0.95	0.90	1.05	1.28	1.18
(5)					0.79	0.67	0.95	1.16	1.07
(6)						0.74	0.41	0.80	0.80
(7)							0.63	0.62	0.61
(8)								0.52	0.53
(9)									0.31

(b) Clustering by the centroid (weighted) strategy

Clustering Cycle	No. of Groups	Clustering Level	Reference SU[a]	SUs in the Group
1	9	0.15	1	2
2	8	0.31	9	10
3	7	0.41	6	8
4	6	0.45	3	4
5	5	0.41	1	2, 3, 4
6	4	0.48	6	8, 9, 10
7	3	0.44	6	7, 8, 9, 10
8	2	0.62	1	2, 3, 4, 5
9	1	0.52	1	All SUs form one group

(c) Clustering by the centroid (unweighted) strategy

Clustering Cycle	No. of Groups	Clustering Level	Reference SU[a]	SUs in the Group
1	9	0.15	1	2
2	8	0.31	9	10
3	7	0.41	6	8
4	6	0.45	3	4
5	5	0.41	1	2, 3, 4
6	4	0.48	6	8, 9, 10
7	3	0.44	6	7, 8, 9, 10
8	2	0.62	1	2, 3, 4, 5
9	1	0.65	1	All SUs form one group

TABLE 16.4 (continued)

(d) Clustering by the group-average strategy				
Clustering Cycle	No. of Groups	Clustering Level	Reference SU[a]	SUs in the Group
1	9	0.15	1	2
2	8	0.31	9	10
3	7	0.41	6	8
4	6	0.45	3	4
5	5	0.56	1	2, 3, 4
6	4	0.61	7	9, 10
7	3	0.67	6	7, 8, 9, 10
8	2	0.79	1	2, 3, 4, 5
9	1	1.13	1	All SUs form one group

(e) Clustering by the flexible strategy with $\beta = -0.25$				
Clustering Cycle	No. of Groups	Clustering Level	Reference SU[a]	SUs in the Group
1	9	0.15	1	2
2	8	0.31	9	10
3	7	0.41	6	8
4	6	0.45	3	4
5	5	0.67	5	7
6	4	0.72	1	2, 3, 4
7	3	0.84	6	8, 9, 10
8	2	1.04	5	6, 7, 8, 9, 10
9	1	1.93	1	All SUs form one group

[a] Lowest numerical value of SUs in group.

these CA strategies will usually give very similar results when somewhat well-defined groups exist in the D matrix. The Wisconsin forest data set illustrates this quite well; because of the simplicity of the data in Table 11.6a, the patterns are obvious and the CA results are consistent with these observations. Where problems arise is when the data sets are large and complex, and with patterns that are not obvious *a priori* to the ecologist. The different strategies may give somewhat different results and, because well-defined cluster patterns may not necessarily emerge in any of the strategies, caution is urged. Note that these CA results differ slightly from the classification of SUs by association analysis (Section 15.4) where SUs 1–5 grouped separately from SUs 6–10 (see Figure 15.2)

16.6   ADDITIONAL TOPICS IN CLUSTER ANALYSIS

In this chapter four hierarchical clustering methods were presented that operate under the combinatorial linear model of Lance and Williams (1967). These methods are computationally efficient once the distance matrix (D) is calculated, since it contains all the information needed to cluster the SUs. Other strategies require the repeated use of the original D matrix at each cycle, which becomes very tedious and is computationally inefficient. Extensive reviews of various CA methods are given in Anderberg (1973), Gauch (1982), Goodall (1978a), Orloci (1978), Pielou (1977, 1984), Romesburg (1984), Sneath and Sokal (1973), and Whittaker (1978b).

Another hierarchical clustering method that is popular with ecologists is *minimum variance clustering* (also known as *Ward's method*—see Everitt 1974, Hartigan 1975, and Orloci 1967a). This method has great intuitive appeal because it is based on the simple underlying principle that at each stage of clustering the variance *within* clusters is minimized with respect to the variance *between* clusters. The within-group variance is defined as the sums of squares of the distances between SUs within the cluster and the centroid of the cluster. At each clustering cycle, the two SU clusters whose fusion results in the smallest (minimum) increase in variance (relative to the variances within each cluster taken separately) are joined. The computations required by this method can be done through the use of SAS (Statistical Analysis Systems, Ray 1982) under the cluster procedure option for Ward's method.

As briefly mentioned in Section 16.2, delimiting homogeneous communities from the information provided by the clustering process is usually done subjectively. The ecologist usually has some feeling about the number of communities (groups of SUs) expected from the given data set, and it is a simple matter to "cut the stems" in the dendrogram (e.g., Figure 16.4) at the clustering level that gives this number of SU groups or communities.

There are, however, numerous objective methods that can be used along with the intuition of the ecologist. One of the earliest methods proposed for evaluating dendrograms was *cophenetic correlation* (Sneath and Sokal 1973), where the distances between SUs implied from the dendrogram are compared to the original SU × SU distance (D) matrix. Proceeding through each cluster cycle, as larger and larger groups are formed (ceasing when all SUs are joined), the correlation between the original D matrix distances and the dendrogram distances will drop. A large drop in this correlation from one cycle to another would suggest stopping fusion at the previous cycle. For example, if the cophenetic correlation for the CA shown in Figure 16.4 was 0.80 at the chord distance 1.0 (upper dashed line) but dropped to 0.50 at distance 1.5 (lower dashed line), then perhaps the clusters formed by cutting the stems at 1.0 should be accepted as homogeneous communities.

Orloci (1967a) suggested that significance levels could be determined for delimiting homogeneous groupings during the minimum variance clustering procedure. Goodall (1978a) formalized an approximate variance ratio for testing whether the increase in variance caused by the fusion of an SU or SU group with another SU group is within an "acceptable" (e.g., $P = 0.05$) level. Ratliff and Pieper (1981) generalized this analysis-of-variance approach to include a test of the hypothesis that the mean intracluster distance is not significantly different from the mean intercluster distance. Their procedure begins with applying the test at the *two-group* level, that is, testing for differences in mean distances between all SU as one group versus being split into two groups. This follows the procedure for Hill's (1980) *stopping rule*, which is a similar approach. For other recent developments in evaluating classifications, the student is referred to the reviews by Archie (1984) and Rohlf (1974) and studies by Duncan and Estabrook (1976), Popma et al. (1983), and Rohlf (1982).

The CA strategies shown in Table 16.1 (centroid, group average, and flexible) are considered *space conserving*; the clustering of SUs at the various levels (distances) introduces relatively little distortion when these clustering distances are compared to the original SU × SU D matrix distances.

We have not included all the CA strategies that operate under the Lance and Williams linear combinatorial model. For example, the single linkage and complete linkage strategies (Sneath and Sokal 1973) are omitted. These two CA methods are strongly *space distorting*, either by contracting distances with "single" linkages or by dilating distances with "complete" linkages (Pielou 1977). That is, after fusion of clusters the reconstructed distance matrix differs greatly from the original distance matrix (D).

The centroid strategies frequently result in what are termed *reversals*: the distance between centroids of some pair may be less than that between another pair merged at an earlier cycle. If the sum of the parameters α_1, α_2, and β equals 1, then successive hierarchical joinings will be monotonic and reversals will not occur. The flexible strategy is, therefore, by definition monotonic since the sum of the parameters is constrained to equal 1. The flexible strategy's chief feature is that by varying β, which controls the space-conserving properties of the clustering strategy, the space can be made to either dilate or contract. A β near -0.25 tends to be space-conserving, but as β becomes more negative, the distortion is toward dilating and as β becomes more positive, the distortion is toward contracting. We suggest the student refer to Sneath and Sokal (1973) for further details on the flexible strategy.

In this chapter we have presented cluster analysis, assuming the data were from SUs randomly dispersed over the landscape and with species abundances or presence–absence observations. However, clustering SU data from a time sequence can be used to examine ecological succession models (Legendre et al. 1985). Other types of observations can also be used for clustering, for example, forest tree size classes or soil profile data (Faith et al. 1985).

16.7 SUMMARY AND RECOMMENDATIONS

1. Cluster analysis (CA) is a technique that accomplishes the sorting of objects (SUs) into groups or clusters based on their overall resemblance to one another. Similar SUs will form clusters distinct from other clusters of SUs. *Cluster analysis* is a general term that refers to a large number of such algorithms that differ mainly in their treatment of cluster formation.

2. The Lance and Williams (1967) general algorithm for CA is a linear combinatorial equation [Eq. (16.1)]. By selecting various values for the parameters of Equation 16.1 (as shown in Table 16.1), four CA strategies can be accomplished: the centroid (both weighted and unweighted), group-average, and flexible.

3. The results of CA are conveniently summarized in a dendrogram (e.g., Figure 16.2). The identification of specific groups or communities from this dendrogram is somewhat subjective. As a general guideline, we recommend not dividing so finely that you end up with a large number of fragmented clusters. Some objective methods have been proposed by several researchers (Section 16.6). Although it is helpful for the student to use methods such as CA as an aid in interpreting data, it is a mistake to place strong emphasis on results of a single analysis.

4. Most of the CA strategies usually give very similar results when the basic data set being analyzed is, in fact, one that is characterized by some relatively obvious patterns (such as the example in Section 16.5). However, in cases when the data set is large, somewhat complex in nature, and with no obvious patterns, the results of the various CA strategies can often vary (in some cases, substantially). In the latter case, we recommend that alternative strategies be explored and their results compared; such comparisons often help identify logical clusters (in view of the underlying ecological knowledge of the data).

PART SIX
COMMUNITY ORDINATION

CHAPTER 17

Background

An area of statistical ecology that has undergone great development in recent years is community ordination. *Ordination* is a term used to describe a set of techniques in which sampling units (SUs) are arranged in relation to one or more coordinate axes such that their relative positions to the axes and to each other provides maximum information about their ecological similarities. By identifying those SUs in a collection that are the most similar (or dissimilar) to one another based on coordinate position, we can then search for underlying factors that might be responsible for the observed patterns. The ultimate goal is, by inference, to elucidate those biological and environmental factors that may be important in determining the structure of the ecological communities from which the samples are drawn. The aim of ordination is to simplify and condense massive data sets in the hope that ecological relationships will emerge. Orloci (1978) discusses this "summarization" aim as one of the major goals of ordination.

Conceptually, an ordination can be visualized as placing SUs within a *species hyperspace*, that is, a hyperspace where there is a single dimension or axis for each species (see Figure 13.2). In the simplest case, assume that a single property, say the abundance of one species, A, has been measured for each SU in a sample. Now, the SUs can be positioned along this single abundance axis (i.e., the abundance of species A from 0 to 100). SUs located at opposite ends of this axis are, of course, highly dissimilar and those located side by side on the axis are similar. This graphic approach can also be extended to the case of two or three species simultaneously, but, obviously, this rapidly becomes intractable, particularly when there are many species and SUs.

The objectives of ordination should be clearly delimited from those of community classification (Chapters 15 and 16). Given an ecological data matrix X with S rows (species) and N columns (SUs), the objective of a classification is to reduce $X_{S,N}$ into $g < N$ "homogeneous" groups or clusters, that is, the SUs within each of the g clusters are more similar to one another than are the SUs between clusters (Green 1980). Ordination, on the other hand, performs a transformation on the $S \times N$ matrix and reduces it to a $p < S \times N$ matrix. This transformation is often referred to as a "reduction of dimensionality" (Legendre and Legendre 1983). Obviously, if we have many species and can reduce the dimensions to only a few (say, $p = 2$), relations between the SUs can more readily be examined. One of the goals of ordination is to minimize the loss of information in deriving these p dimensions (Orloci 1974a).

Goodall (1954b) introduced the term *ordination* to identify a series of techniques used for arranging vegetation samples (or species) in relation to "a multidimensional series" (Austin 1985). Goodall's work was prompted by the need to quantify the concept of vegetation as a continuum, that is, vegetation samples or SUs represent a sample from a "continuous" real world (the *continuum* and *individualistic* concepts—Gleason 1926, McIntosh 1967). These two complementary concepts contrast with the discrete, *organismic concept* of biotic communities that underlies classification (although some view this contrast as a "nonexistent" problem—see Anderson 1965). Although the abstract patterns that can emerge from an ordination may suggest discrete communities (i.e., groups of SUs within the coordinate system), the product of an ordination is usually not one of classification. However, a seemingly natural and useful community classification may "secondarily" follow from the results of an ordination (see Jensen 1979) and ordination can be applied to subsets of large data sets split by classification.

Two methods of ordination can be distinguished: (1) *direct*, and (2) *indirect* (Gauch 1982, Whittaker 1967). In a direct ordination, SUs are positioned along measured environmental gradients that were selected beforehand as the basis for the study (Austin et al. 1984). Although we do not cover these direct approaches, a brief overview is given in Chapter 20. With indirect ordination, the SUs are arranged within a reduced coordinate system based on their dissimilarities (or similarities) in species composition; thus, SUs are related via synthetic axes or gradients and the ordination is considered indirect (Whittaker 1967, 1978a). Several indirect ordination methods are described in Chapters 18 to 21.

Historically, in addition to Goodall's (1954b) work, early development of techniques for ordination in community ecology can be traced back to the pioneering work by the Wisconsin plant ecologist John Curtis and his students (e.g., Brown and Curtis 1952, Curtis and McIntosh 1951). Dissatisfaction with some of the early single-axis ordinations logically lead to the development of a multiple-axis method, published by Bray and Curtis in 1957. This method has popularly been referred to as *Bray–Curtis* or *Wisconsin ordination*; it is now generally referred to as *polar ordination* (see Beals 1984, Gauch 1982, Orloci 1978). The procedures for a polar ordination are described in Chapter 18.

Another multiaxis ordination method, first used in ecology by Goodall (1954b), is *principal components analysis* (PCA). Initially Goodall's paper on PCA (which he called *factor analysis*) largely went unnoticed, but as computers became generally available, PCA became a widely used ordination method in ecology. PCA is described in Chapter 19. However, PCA has several limitations, particularly when dealing with community data sets that have strong nonlinearities. Correspondence analysis, a technique that is somewhat more robust with nonlinear data sets than PCA, is described in Chapter 20. Finally, two nonlinear methods, polynomial ordination and nonmetric multi-dimensional scaling, are discussed in Chapter 21.

Austin (1985), Beals (1984), Gauch (1982), and Whittaker (1978a) provide excellent discussions on the concepts and purposes of ordination. We recommend these references as a place to start for those students interested in those aspects of ordination beyond the scope of this book.

Whittaker and Gauch (1973) listed the following properties of ordination procedures, which may also be used as evaluation criteria: (1) relative freedom from distortion, (2) lucidity or clarity of results, (3) efficient use of data, and (4) heuristic value for revealing otherwise unrecognized patterns and relationships. We will use these properties to evaluate the different ordination methods in Chapters 18–21.

17.1 MATRIX VIEW

The ordination process begins by examining the total set of relationships between species and SUs within the community data matrix (Figure 17.1).

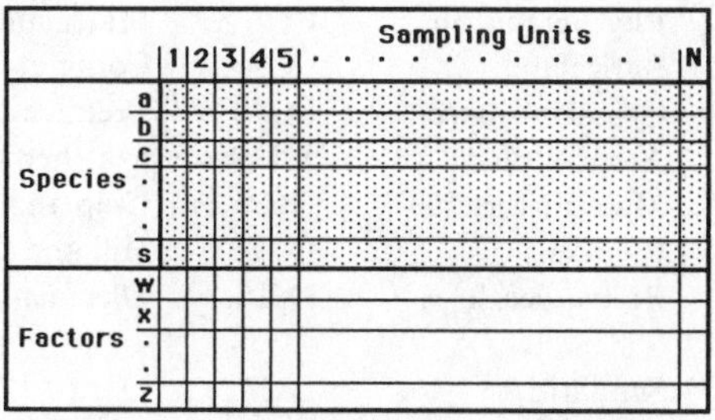

Figure 17.1 *The shaded area indicates the form of the ecological data matrix for ordination. SU ordination may be based on R-mode resemblances between species (i.e., by rows) or Q-mode resemblances between SUs (i.e., by columns).*

Depending on the particular method used, resemblances are calculated between pairs of SUs or pairs of species. It is from these SU or species pairwise comparison matrices that the coordinate system is derived. Recall that the aim of ordination is to derive axes systems showing the relationships between SUs. If the SU ordination is based on R-mode resemblances between species (i.e., by rows), it is referred to as an *R-strategy*. An SU ordination computed using *Q*-mode resemblances between SUs (i.e., by columns) is referred to as a *Q-strategy* (Noy-Meir and Whittaker 1978, Orloci 1978).

17.2   SELECTED LITERATURE

The use of ordination methods by ecologists shows a natural progression from the early use of polar ordinations to the more recent emphasis on principal components analysis, detrended correspondence analysis, and nonmetric multidimensional scaling (Table 17.1). When the ecological data matrix has non-

TABLE 17.1   *Selected literature for examples of ecological ordination studies using polar ordination (PO), principal components analysis (PCA), correspondence analysis/detrended correspondence analysis (DCA), or nonmetric multidimensional scaling (MDS).*

Location	Community	Method	Reference
Wisconsin	Upland forest	PO	Bray and Curtis 1957
Wisconsin	Prairie	PO	Dix and Butler 1960
Wisconsin	Forest birds	PO	Beals 1960
Utah	Benthos	PO	Erman 1973
England	Chalk grassland	PCA	Austin 1968
Ontario, Canada	Spruce forest	PCA	Jeglum et al. 1971
Sask., Canada	Forest lichens	PCA	Jesberger and Sheard 1973
Ghana	Ants and plants	PCA	Majer 1976
Texas	Fire ants	PCA	Pimm and Bartell 1980
N. Atlantic	Phytoplankton	PCA	Matta and Marshall 1984
Australia	Rangelands	PCA	Foran et al. 1986
Hoy, Orkney	Grassland-herb	MDS	Prentice 1977
Norway	Arctic-alpine	MDS	Matthews 1978
Alberta, Canada	Macroinvertebr.	DCA	Culp and Davies 1980
England	Rich-fens	DCA	Hill and Gauch 1980
Finland	Boreal heath	DCA, MDS	Oksanen 1983
Southeast Spain	Semiarid	DCA, MDS	Dargie 1984
Wales	Dune grassland	DCA	Gibson and Greig-Smith 1986

linear relationships between species, we recommend the use of ordination methods that either: (1) attempt to "correct" for nonlinearities (e.g., detrended correspondence analysis—Hill and Gauch 1980), or (2) circumvent the linearity assumption (e.g., nonmetric multidimensional scaling—Kenkel and Orloci 1986).

CHAPTER 18

Polar Ordination

Polar ordination (PO) was one of the first ordination methods to be widely used by plant ecologists. Although PO has some limitations and more mathematically sophisticated ordination methods are now being used (Chapters 19–21), PO continues to produce ecologically meaningful results. In fact, PO is often used as a standard for comparisons with the newer methods. The Bray and Curtis (1957) method of PO is described in this chapter.

18.1 GENERAL APPROACH

The PO method of Bray and Curtis (1957) is one of the few techniques that was specifically developed to analyze plant community data (Beals 1984). The aim of PO is to position SUs within a coordinate system (axes) such that the distances between SUs reflect their general similarity and, it is hoped, their relation to underlying environmental gradients. Basically, the procedure involves the selection of SUs as endpoints (poles) on an axis, followed by a simple geometric positioning of the remaining SUs relative to these endpoint SUs. The ecologist often has a strong intuitive feel for the environmental differences between the endpoint SUs and, thus, for the gradients represented by the axes.

One feature that has undoubtedly contributed to the appeal of PO is simplicity. The computational steps in PO are straightforward and therefore tend to be less intimidating than more sophisticated mathematical ordination procedures (e.g., principal components analysis).

18.2 PROCEDURES

The procedures outlined below closely follow the original Bray and Curtis method as described by Cottam et al. (1978).

18.2.1 Bray–Curtis Polar Ordination

It is assumed that there are N SUs and that the property of interest in each SU is some measure of abundance (e.g., density, percentage cover, biomass) for each of S species.

STEP 1. COMPUTE SU RESEMBLANCES. As was the case in the classification procedure described in Chapter 16, the first step in PO is to compute the Q-mode resemblances between pairs of SUs. Bray and Curtis used percent similarity [PS, Eq. (14.6a)] as their measure of resemblance. [PS is also known as *Czekanowski's index* (Bloom 1981, Goodall 1978b).] Recall from Chapter 14 that the dissimilarity complement of PS (i.e., $100 - PS$) is termed *percent dissimilarity* [PD, Eq. (14.6b)]. Although Orloci (1978) does not recommend PD for use in *Euclidean-type* ordinations (i.e., those requiring strictly geometric procedures), we have retained its use here because of its traditional application in PO. There are strong advocates for (Beals 1984) and against (Orloci 1974b, 1978) the use of PD in PO. We discuss this topic further in Section 18.6.

For each pair of SUs (j, k), compute their percent dissimilarity (PD) as illustrated in Chapter 14 [Eq. (14.6b)].

STEP 2. CONSTRUCT THE SU × SU RESEMBLANCE MATRIX. The PS and PD values between all pairs of SUs are summarized in SU × SU matrix form with the PS values above the diagonal and the PD values below the diagonal.

STEP 3. SELECT ENDPOINTS AX AND BX FOR THE X AXIS. For each SU, the sum of dissimilarities (PD) with all other SUs is computed.

(a) The SU with the largest sum of dissimilarities is designated AX.
(b) The SU having the largest PD to AX is designated BX. (In selecting BX, if more than one SU has a PD with AX that is greater than or equal to 95%, select the SU with the largest sum of dissimilarities.)
(c) The length (L) of the X axis is given by the PD between AX and BX (i.e., $PD_{AX, BX}$).

STEP 4. POSITION REMAINING SUs ON THE X AXIS. The locations (x) of the remaining SUs on the X axis, relative to the endpoint SUs, AX and BX, are computed using Beals' (1965) formula

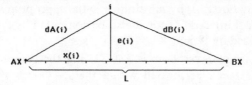

Figure 18.1   *Geometric representation of Beals' (1965) formula for positioning the ith SU on the X axis (defined by endpoint SUs AX and BX) in polar ordination. See Eqs. (18.1) and (18.2) for definition of symbols.*

$$x(i) = \frac{L^2 + dA(i)^2 - dB(i)^2}{2L} \tag{18.1}$$

where $x(i)$ is the location of the ith SU along the X axis, L is the PD between AX and BX, $dA(i)$ is the PD of the ith SU to AX, and $dB(i)$ is the PD of the ith SU to BX. Equation (18.1) is diagrammatically shown in Figure 18.1.

STEP 5. SELECT ENDPOINTS AY AND BY FOR THE Y AXIS.   Note that the distance $e(i)$ from point i to the X axis in Figure 18.1 is given by the Pythagorean theorem:

$$e(i) = \sqrt{dA(i)^2 - x(i)^2} \tag{18.2}$$

It can also be seen from Figure 18.1 that the largest $e(i)$ value will be for an SU that is equally dissimilar to both endpoint SUs, resulting in a tendency for the SU to be equidistant from both AX and BX (i.e., midaxis). To position the SUs on the Y axis, with regard to the greatest component of the variability that remains, $e(i)$ is computed for each SU.

(a)   The SU with the largest $e(i)$ is designated AY. An additional condition for selecting AY is that it must fall within the mid 50% range of the X axis, that is, between $0.25L$ and $0.75L$.

(b)   The SU having the largest PD with AY is selected as BY. A further condition for the selection of BY is that it must lie within 10% of AY on the X axis. This ensures that the Y axis will be approximately perpendicular to the X axis. [Again, if more than one SU has a PD to AY greater than or equal to 95% (and otherwise meets the above conditions), the SU with the largest sum of PD is selected as BY.]

(c)   The length (L) of the Y axis is the PD between AY and BY (i.e., $PD_{AY,BY}$).

STEP 6. POSITION REMAINING SUs ON THE Y AXIS.   The locations (y) of the remaining SUs on the Y axis, relative to the endpoint SUs AY and BY, are computed using Eq. (18.1) by substituting in the appropriate Y-axis endpoints and length. [A third axis (Z) could also be computed by following the same steps as described for X and Y.]

STEP 7. GRAPHICAL DISPLAY OF PO. Using the computed $x(i)$ and $y(i)$ values, each SU can be located within the $X-Y$ axis system, producing a two-dimensional PO.

18.2.2 Simplified Polar Ordination

The early applications of PO were with data sets obtained from relatively homogeneous biotic communities, such as the upland forests of Wisconsin (Bray and Curtis 1957). However, ecologists frequently deal with data sets collected over many different types of environments, for example, the upland and lowland forests in Saskatchewan studied by Swan and Dix (1966). When PO is applied to such heterogeneous data, a serious problem can arise with regard to the selection of the endpoint SUs: a number of SU pairs can be 100% dissimilar and, most likely, the different pairs will be related to different underlying environmental gradients. Thus, which pair of SUs should be selected for endpoints on the first axis?

In cases where the ecologist has a reasonable *a priori* knowledge about the existence of environmental gradients that might explain patterns of species variations in communities, Cottam et al. (1978) and Gauch (1982) argue that endpoint selection for each axis in the PO should be done subjectively. By choosing those SUs that represent the "poles" of the environmental gradients, the problem of ending up with two dissimilar SUs being positioned close together can be reduced. There is an additional computational advantage in that the entire SU dissimilarity comparison matrix is not needed, only the dissimilarities between the chosen two endpoint SUs and the other SUs. In the computational example in Section 18.3, the traditional Bray–Curtis PO calculations as described in Section 18.2.1 will be illustrated, but, in the computer example in Section 18.4, the simplified PO procedure will be used. The advantage of using a computer program with simplified PO is that various "trial-and-error" choices for endpoints can be run and the ordinations quickly examined for the one with the most lucid results and highest heuristic value.

18.3 EXAMPLE: CALCULATIONS

The following calculations illustrate the procedure for a PO using a data matrix for the abundances of three species in five SUs, along with the sums of species abundances in each SU (Table 18.1). Although many studies using PO apply a *double standardization* to the data before commencing, as discussed

TABLE 18.1   *Species (Spp) abundances and means in five sampling units (SUs) and sums for each SU*

Spp	(1)	(2)	(3)	(4)	(5)	Mean
			SUs			
(1)	2	5	5	3	0	3.0
(2)	0	3	4	2	1	2.0
(3)	2	0	1	0	2	1.0
Sums	4	8	10	5	3	

by Beals (1984) and Cottam et al. (1978), we illustrate the PO calculations using untransformed data. Double standardization involves first expressing the abundances for each species relative to the maximum abundance for each species (i.e., *row* standardization) followed by a division of each transformed value by the total of such values for each SU (i.e., *column* standardization). This has many effects, but mainly avoids a strong weighting by a few highly abundant species in the PO calculations (Beals 1984).

STEP 1. COMPUTE SU RESEMBLANCES.   The percent similarity [PS, Eq. (14.6a)] and percent dissimilarity [PD, Eq. (14.6b)] between the (j, k) pairs of SUs are

$$PS(1, 2) = [(2)(2 + 0 + 0)/(4 + 8)](100) = (4/12)(100) = 33\%$$

$$PS(1, 3) = [(2)(2 + 0 + 1)/(4 + 10)](100) = (6/14)(100) = 43\%$$

$$\vdots \qquad\qquad \vdots \qquad\qquad\qquad \vdots \qquad \vdots$$

$$PS(4, 5) = [(2)(0 + 1 + 0)/(5 + 3)](100) = (2/8)(100) = 25\%$$

and

$$PD(1, 2) = 100 - 33 = 67\%$$

$$PD(1, 3) = 100 - 46 = 57\%$$

$$\vdots \qquad\qquad \vdots \qquad\quad \vdots$$

$$PD(4, 5) = 100 - 25 = 75\%$$

STEP 2. CONSTRUCT THE SU × SU RESEMBLANCE MATRIX.   The percent similarity–dissimilarity matrix, along with the sum of PD for each SU, is given in Table 18.2.

TABLE 18.2 *Matrix of PS (above diagonal) and PD (below diagonal) values between the five SUs. The sum of PD for each SU is also given*

		PS—Percent Similarity				
	SUs	(1)	(2)	(3)	(4)	(5)
	(1)	—	33	43	44	57
	(2)	67	—	89	77	18
PD—Percent	(3)	57	11	—	67	31
Dissimilarity	(4)	56	23	33	—	25
	(5)	43	82	69	75	—
	Sum PD	223	183	170	187	269

STEP 3. SELECT ENDPOINTS AX AND BX ON THE X AXIS. SU 5 has the largest sum of dissimilarities (PD = 269) and is designated as endpoint AX. SU 2 has the greatest dissimilarity with SU 5 (PD = 82%) and is thus selected as endpoint BX. The length (L) of the X axis is 82.

STEP 4. POSITION REMAINING SUs ON THE X AXIS. Using Eq. (18.1) with $L = 82$, the location of SU 1 on the X axis is

$$x(1) = [(82)^2 + (43)^2 - (67)^2]/(2)(82)$$

$$= (6724 + 1849 - 4489)/164 = 4084/164 = 25$$

and, similarly, for SUs 3 and 4

$$x(3) = 69$$

$$x(4) = 72$$

STEP 5. SELECT ENDPOINTS AY AND BY FOR THE Y AXIS. The first step is to compute the $e(i)$ values for the remaining SUs using Eq. (18.2). For SU 1,

$$e(1) = \sqrt{[(43)^2 - (25)^2]} = \sqrt{1224} = 35$$

and, similarly, for SUs 3 and 4

$$e(3) = 0$$

$$e(4) = 21$$

SU1, with the largest $e(i)$ value and lying on the X axis at 25 (which is within

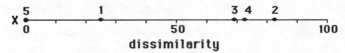

Figure 18.2 Single-axis Bray–Curtis PO for five SUs, based on percentage dissimilarity.

the mid 50% range of the X axis) is selected as the AY endpoint. However, there is no other SU that lies within 10% of AY and, consequently, the ordination is restricted to a single axis in this simple example.

STEP 6. GRAPHIC DISPLAY OF PO.   The results of the single-axis PO are shown in Figure 18.2. Note that the relative positions of SUs along the X axis are very similar to the order of clustering of these same SUs by cluster analysis (Figure 16.2).

18.4 EXAMPLE: PANAMANIAN COCKROACHES

The simplified PO procedure (Section 18.2.2) was applied to data for the abundances of five cockroaches in six Panamanian localities (Table 11.4a). The BASIC program PO.BAS (see accompanying disk) was used with PD as the resemblance measure, although PO.BAS provides options for other distance measures.

Note, from Table 18.3, that PO.BAS prompts the user for choices of X and Y axis endpoints. From Table 11.4a we find that SU 6 shares no common species with either SU 3 or SU 4. Thus, it seems reasonable to try each of these combinations (i.e., SUs 6–3 and 6–4) as endpoints for the X axis. The choice of SUs 6 and 4 gave us the clearest separation of SUs on the X axis (Figure 18.3). For the Y axis, we tried several of the remaining combinations and arrived at SUs 1 and 5 as providing the most lucid result. (For this particular example, this "trial-and-error" approach gave identical results to those using the full Bray–Curtis PO procedures.)

In Figure 18.3 the 6 SUs are located in the space of a two-axis coordinate system, which reflects their overall similarity, and we have achieved "reduced dimensionality" since the original data matrix of 5 species and 6 SUs has now been transformed to a 2×6 matrix. Although this ordination reflects the abundances of *all* species, the dominant species (*Latindia dohrniana*) has a large influence on the PO. It has high abundance in the SUs at one end of the X axis (SUs 1, 2, 5 and 6) and its abundance is 0 in SUs 3 and 4 at the other end of X (Figure 18.3).

TABLE 18.3 *Polar ordination results as summarized for six Panamanian localities using BASIC program PO.BAS*

Enter the SU numbers for endpoints of the X axis? 4, 6
Enter the SU numbers for endpoints of the Y axis? 1, 5
Percentage dissimilarity between endpoint SU 4) and SU

| 1) = 98.3 | 2) = 97.2 | 3) = 33.3 | 5) = 88.2 | 6) = 100.0 |

Percentage dissimilarity between endpoint SU 6) and SU

| 1) = 66.2 | 2) = 50.0 | 3) = 100.0 | 4) = 100.0 | 5) = 46.3 |

Percentage dissimilarity between endpoint SU 1) and SU

| 2) = 50.0 | 3) = 96.6 | 4) = 98.3 | 5) = 75.9 | 6) = 66.2 |

Percentage dissimilarity between endpoint SU 5) and SU

| 1) = 75.9 | 2) = 63.2 | 3) = 77.8 | 4) = 88.2 | 6) = 46.3 |

X and Y coordinates for a two-dimensional polar ordination

SU	X	Y
(1)	76.4	0.0
(2)	84.8	28.1
(3)	5.6	59.6
(4)	0.0	50.3
(5)	78.2	75.9
(6)	100.0	52.6

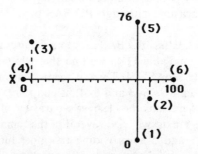

Figure 18.3 *Polar ordination of six Panamanian localities based on abundances of five cockroach species. The X and Y axes are drawn between their respective endpoint SUs to emphasize the concept of these SUs as "poles" for each axis.*

Of course, in the final analysis, the value of any PO lies in its heuristic value. Does it provide a useful ecological interpretation? Do the X and Y axes reflect patterns in species abundances and do they reflect underlying environmental factors? These important questions, and the methods for addressing them, are described in Chapter 24.

TABLE 18.4   *Polar ordination results: (a) PD matrix and sums, and (b) X coordinates and deviations for each of the 10 forest sites (SUs), using program PO.BAS.*

(a) Percent dissimilarity matrix and sums

SUs	(1)	(2)	(3)	(4)	(5)	(6)	(7)	(8)	(9)	(10)
(1)	—	7.4	29.4	25.4	43.9	70.4	56.9	71.9	81.5	75.4
(2)	7.4	—	33.3	25.4	43.9	70.4	53.8	71.9	77.8	75.4
(3)	29.4	33.3	—	17.9	48.1	72.5	64.5	77.8	92.2	77.8
(4)	25.4	25.4	17.9	—	25.8	55.9	42.9	61.3	79.7	67.7
(5)	43.9	43.9	48.1	25.8	—	40.4	32.4	53.3	71.9	60.0
(6)	70.4	70.4	72.4	55.9	40.4	—	35.4	19.3	40.7	43.9
(7)	56.9	53.8	64.5	42.9	32.4	35.4	—	29.4	32.3	26.5
(8)	71.9	71.9	77.8	61.3	53.5	19.3	29.4	—	22.8	23.3
(9)	81.5	77.8	92.2	79.7	71.9	40.9	32.3	22.8	—	15.8
(10)	75.4	75.4	77.8	67.7	60.0	43.9	26.5	23.3	15.8	—
Sum	462.2	459.4	513.5	402.0	419.6	448.9	374.1	431.1	514.6	465.8

(b) Coordinates and deviations for SUs on X axis

SUs:	(1)	(2)	(3)	(4)	(5)	(6)	(7)	(8)	(9)	(10)
$x(i)$:	77.4	72.9	92.2	78.8	61.6	26.5	29.2	16.1	0.0	14.6
$e(i)$:	25.4	27.2	0.0	11.8	37.2	30.9	13.9	16.2	0.0	6.0

18.5   EXAMPLE: WISCONSIN FORESTS

Using the abundances for eight trees in 10 upland forest sites in southern Wisconsin (Table 11.6a), the BASIC program PO.BAS was again used to compute a PO (Table 18.4). SU 9 has the largest sum of PD with the other SUs (514.6) and is selected as *AX*. SU 3 has the greatest PD with SU 9 (92.2) and is selected as *BX*.

After locating the coordinates for the remaining SUs along the *X* axis, the deviation (*e*) of each SU from the *X* axis is computed (Table 18.4*b*). SU 5 has the largest deviation from *X* (37.2), falls within the midrange of *X*, and, hence, is the first endpoint for the *Y* axis (*AY*). However, there are no SUs within 10% of SU 5 and, consequently, a *Y* axis is not computed, since our criteria for endpoints on the *Y* axis cannot be satisfied. (Note that SU 2 is within 12% of SU 5, but its use as endpoint *BY* would result is an unacceptable oblique angle between the *Y* axis and the *X* axis—unacceptable because for *X* and *Y* to represent independent environmental gradients, they should be approximately orthogonal or perpendicular.)

A plot of the 10 forest sites along the PO *X* axis shows that SUs 1, 2, and 4 are located near the endpoint SU 3, indicating their close resemblance with

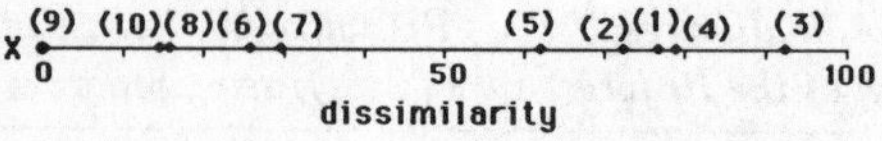

Figure 18.4   *Single-axis (X) polar ordination of 10 upland forest sites, Wisconsin.*

SU 3 (Figure 18.4). These SUs are dominated by bur oak and black oak. SUs 6, 7, 8, and 10 (dominated by sugar maple and basswood) are located near the pole defined by SU 9, thus are similar to SU 9.

18.6   ADDITIONAL TOPICS IN POLAR ORDINATION

Sneath and Sokal (1973) point out the advantages of using a Q-mode resemblance measure, which permits the relationships between SUs to be visualized geometrically. While this geometric space need not be Euclidean, it should be *metric* so that the topology of the space is determined solely by the resemblance function. One condition a metric satisfies is known as the triangle inequality condition

$$D(j, k) < D(j, l) + D(k, l) \tag{18.3}$$

The percent dissimilarity (PD) measure we used for PO in this chapter fails this condition under certain situations. For example, from Table 18.2, $D(2, 5) = 82$, $D(2, 3) = 11$, and $D(3, 5) = 69$ and, thus, $D(2, 5) > D(2, 3) + D(3, 5)$ or $82 > 11 + 69$, which fails Eq. (18.3). Because of the importance of geometry in PO, Orloci (1974b) recommends the exclusive use of metric resemblance measures. However, Beals (1984) points out that the triangle inequality problem with PD disappears if the data are standardized (see below). The BASIC program PO.BAS provides options for several distance functions, and we encourage the student to run the program with PD and one or two other measures and examine the *ecological consistency* of the results. If inconsistent, this suggests that it may be worth investigating whether the differences are due to the measure, the data, or a combination of both.

Orloci (1974b, 1978) illustrates the nonmetric property of PD and shows that its underlying cause is in the divisor of PS [Eq. (14.6a)], which is the sum of all observations for a pair of SUs. Since this sum could be different for each pair of SUs, each PS value can be on a different scale (different divisor) and, hence, the loss of the metric property. Beals (1984) notes that if the data are *column standardized*, that is, each species abundance for a given SU is divided by the sum for the SU, then the divisor is a constant and PS (and its complement PD) will be metric. Furthermore, Beals states that "all the evidence in the literature shows that the Sorensen coefficient (PS) gives ecologically more interpretable results in multivariate analysis" than many metric measures, such as Euclidean distance.

Some of the statistical properties of PS are known. Ricklefs and Lau (1980) derived by computer simulation the dispersion (standard deviation) characteristics of PS in relation to different sample sizes. As expected, standard deviations were greatest at small sample sizes. They also found the standard deviation to be smallest when the similarity between SUs was either low (near 0%) or high (near 100%).

As shown by the Wisconsin forest example (Section 18.5), a problem with polar ordination is that the Y axis can be oblique to the X axis because the "best" endpoint SUs for the Y axis are often some distance apart on the X axis. Recall that if Y is oblique to X, these ordination axes cannot represent completely independent gradients. This oblique problem is less common for larger data sets where there are more choices of endpoint SUs for the Y axis. Orloci (1974b, 1978) provides the details for both an analytical and a graphic solution for correcting for nonperpendicular PO axes.

Another problem confronting the ecologist is very heterogeneous data. An inherent characteristic of such data is that species-abundance relationships tend to be strongly nonlinear (this is often termed *nonlinear data structure*). All ordination methods are affected to some degree by such nonlinearities (Kenkel and Orloci 1986). Although the Bray–Curtis PO is perhaps less sensitive to nonlinear data structures than some other ordinations (Gauch and Whittaker 1972), Cottam et al. (1978) have noted that the Wisconsin polar ordination is "by no means a fool-proof technique" for dealing with nonlinearities. Nonlinearity is a major problem in ordination theory and methodology, but we defer this problem to Chapter 21.

These two problems (endpoint SU selection and nonlinearity), and other less serious problems, have resulted in numerous modifications to the original Bray–Curtis PO (Bannister 1968, Beals 1984, Gauch and Scruggs 1979, Orloci 1974b). Ecologists have modified the fine workings of the basic technique to suit the peculiarities of their data. Although this opens up ecologists to criticism for being subjective, we feel that ecologists best knows their own data and how to best extract meaningful ecological patterns from them (using PO, in this case). We encourage the student to modify the PO criteria for selecting endpoints along lines suggested by Beals (1984), particularly when dealing with very heterogeneous community data. The heterogeneity problem is also related to the selection of endpoint SUs—be aware that in selecting an "outlier" SU as an endpoint the other SUs tend to bunch at the end of the axis away from the outlier SU (Newsome and Dix 1968).

18.7 SUMMARY AND RECOMMENDATIONS

1. Polar ordination (PO) has appeal because of its procedural and conceptual simplicity relative to more mathematical ordination methods. Although often criticized for its subjectivity, we agree with Beals (1984) that PO is a valuable method.

2. During each step of PO, rather subjective decisions are sometimes required (e.g., should data be transformed? which dissimilarity measure? which endpoint SUs?). However, probably one of the main reasons PO has been so successfully used in ecological studies is that the ecologist is directly involved in the straightforward procedural decisions required by the method. Some of the other ordination methods tend to be "black boxes" where data are fed into a computer program and the user is largely detached from the analysis.

3. An important step in PO is the selection of a Q-mode resemblance measure between pairs of SUs (Section 18.2.1). Selecting an appropriate resemblance measure for each data set is sometimes difficult. Orloci (1974b) recommends metric measures (e.g., chord distance) that give importance to the "relative" differences in species abundance between SU pairs (Section 18.6). Beals (1984) recommends using PS. We recommend that several measures and data standardizations be tried and the results be compared. If there are fairly obvious patterns in the data, these will consistently show up in each analysis.

4. At the critical step of endpoint selection, the ecologist must establish selection criteria that will result in ecologically meaningful axes. Although, he/she may be criticised for being subjective, we recommend that the ecologist always use the best *a priori* knowledge available when selecting endpoints to reflect underlying gradients.

5. We recommend that the student avoid the selection of "outlier" endpoint SUs, which can cause the other SUs to be bunched at the opposite end of the axis. Gauch (1982) and Beals (1984) provide in-depth treatments of problems associated with endpoint selection.

6. For very heterogeneous community data, strong nonlinear species-abundance relations can severely distort SU patterns that emerge in PO (or any other ordination method; Kenkel and Orloci 1986). In such cases, we recommend that the data be subdivided into more homogeneous subsets (possibly by classification methods) or that the ecologist apply a nonlinear ordination method (Chapter 21). If PO is used, be aware of the types of distortions that can occur.

CHAPTER 19

Principal Components Analysis

Principal components analysis (PCA) was, for a time, perhaps the most widely used ordination procedure in ecology. Its initial appeal was based on its apparent (and perhaps a bit transparent) mathematical elegance and, more important, its availability in computer statistical packages. The ecologist could simply "plug" his or her data into one of these packages and let the computer program compute the coordinates for an SU ordination—no difficult decisions were needed as to which resemblance measures were appropriate or how to select endpoint SUs (as in polar ordination, see Chapter 18). With time, problems with PCA as an ordination technique became evident and the search for new techniques was on. However, when used within the range of its intended limits, PCA remains a valuable ordination method. We present the method of PCA in this chapter, along with a discussion of some of its strengths and weaknesses.

19.1 GENERAL APPROACH

PCA was first used in ecology by Goodall (1954b), but its development can be traced back to Pearson's (1901) paper "on lines and planes of closest fit to systems of points in space." PCA is basically a multivariate statistical technique that deals with the internal structure of matrices. It is our experience that the technical aspects of PCA are best understood after the student has worked through some examples. A brief, somewhat technical, overview of PCA is given below, which we recommend the student reread after working through later sections (19.2–19.5). The rich jargon of PCA is definitely less intimidating after working through an example!

PCA is a method of breaking down or partitioning a resemblance matrix into a set of orthogonal (perpendicular) axes or components. Traditionally, this matrix consists of variance–covariances or correlations, but PCA can also be applied to matrices of Euclidean distances (Gower 1966, Orloci 1973). Each PCA axis corresponds to an eigenvalue of the matrix. The eigenvalue is the variance accounted for by that axis. PCA is often referred to as one of the eigenanalysis ordination methods (Orloci 1978, Pielou 1984).

In PCA, the eigenvalues of the resemblance matrix are extracted in descending order of magnitude such that the corresponding PCA axes (components) represent successively greater to lesser amounts of variation in the matrix. Hence, the first few PCA axes, upon which the SUs will be positioned, will represent the largest percentage of the total variation that can be explained (see discussion by Gauch 1982). The result is a reduced coordinate system that provides information about the ecological resemblances between SUs.

Problems associated with the use of PCA in community ecology, which we alluded to in the previous chapter, stem from the fact that PCA is a *linear* model, that is, the coordinates of an SU in the space of the PCA axes system are determined by a linear combination of weighted species abundances. If nonlinear relationships exist, as is often the case in ecological data, a linear ordination model such as PCA will poorly represent the true SU relationships. Orloci (1974a) terms this *type A distortion*. In Chapter 21 we discuss nonlinear ordination models developed to reduce such distortion. Gauch and Whittaker (1972) and Noy-Meir and Whittaker (1977) suggest that PCA might best be applied to data obtained from a relatively narrow range of environmental and compositional variation where, presumably, the linear model will apply.

19.2 PROCEDURE

The method of PCA we outline below is adopted from Austin and Orloci (1966) and Orloci (1966, 1967b, 1973, 1978). Of the numerous ways in which a PCA can be computed, we think Orloci's strategy is the most pedagogic. The strategy is diagrammed in Figure 19.1 and the computational steps are detailed below. We encourage the student to simultaneously work step by step through the procedures in this section with those in Section 19.3 (example calculations).

STEP 1. STANDARDIZE THE DATA MATRIX.   As a reminder, recall that the community data matrix X (see Figure 17.1) has S rows (species) and N columns (SUs). In matrix notation this is written as $X_{S \times N}$, where x_{ij} is the observation (e.g., density, biomass) in the ith row (species) and jth column (SU). The first step in the Orloci strategy is to standardize this X matrix by replacing each

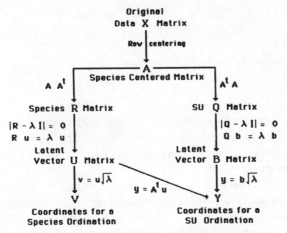

Figure 19.1   A strategy for the R/Q duality (species and SUs) in PCA ordination. Based on papers by Orloci (1966, 1967b, 1973, and 1978).

x_{ij} element with

$$a_{ij} = \frac{x_{ij} - \bar{x}_i}{F_i} \tag{19.1}$$

where $\bar{x}_i$ is the mean value of the ith row (species) and F_i is a standardization function. In the simplest case, if we choose $F_i = 1$ in Eq. (19.1), this produces what is termed *row centering*, that is, a move of the origin of the original coordinate system to the centroid (center of gravity) of the *species space* (see below and Orloci 1974a). However, here we select the standardization function as

$$F_i = \sqrt{\sum_{j=1}^{N} (x_{ij} - \bar{x}_i)^2} \tag{19.2}$$

that is, the square root of the sum of squared deviations about the row mean. Using Eq. (19.1), a new matrix $A_{S \times N}$ is formed with elements a_{ij}.

STEP 2. COMPUTE SPECIES AND SU SIMILARITIES.   R-mode (species) resemblances are now computed by postmultiplying A by its transpose:

$$R_{S \times S} = A_{S \times N} A_{N \times S}^t \tag{19.3}$$

where R is a matrix of r_{ij} *Pearson product-moment correlations* between all

i, j species pairs and A^t is the transpose (rows and columns switched) of A. The correlation coefficients arise from Eq. (19.3), owing to our specific choice of F_i in Eq. (19.2).

Q-mode (SU) resemblances are given by

$$Q_{N \times N} = A^t_{N \times S} A_{S \times N} \tag{19.4}$$

where Q is a matrix of q_{ij} *scalar product similarities* between all ith and jth SUs. [For the student unfamiliar with matrix multiplication, we recommend reviewing the examples given in Searle (1966) and Pielou (1984), or carefully working through the example in Section 19.3.]

It turns out that the R and Q matrices, as derived above, contain essentially the same information (Noy-Meir and Whittaker 1977). Hence, the remaining steps in the SU ordination below could be followed using *either* matrix to obtain identical results. This is illustrated in Figure 19.1. If the number of SUs in the sample (N) is much larger than the number of species (S) (the usual case in ecological studies), then an *R-strategy* would be most computationally efficient (i.e., a PCA on the R matrix). However, if S is much larger than N, a *Q-strategy* would be computationally preferred (i.e., a PCA on the Q matrix). An R-strategy is described below, but, from Figure 19.1, the student should recognize that the mathematical steps in a Q-strategy are the same.

STEP 3. *COMPUTE EIGENVALUES AND EIGENVECTORS OF R.*   A detailed description of the methodologies involved in computing eigenvalues and their associated eigenvectors for matrices is beyond the scope of this book. There are numerous computer programs that can perform these analyses. On the other hand, it is important for the student to have at least an appreciation for these computations before relying on "canned" programs. Thus, an overview is given below.

The eigenvalues (also known as *latent roots*) of the $R_{S \times S}$ matrix, denoted by the Greek letter λ, are obtained by solving the equation

$$|R_{S \times S} - \lambda I_{S \times S}| = 0 \tag{19.5}$$

where I is the identity matrix (1's down the principal diagonal and 0's elsewhere). Equation (19.5) is known as the *characteristic equation*, a polynomial in λ of degree S. The vertical bars indicate that the determinant of the resulting matrix is to be computed (again, the student unfamiliar with any of these terms can refer to the example in Section 19.3). For an $S \times S$ matrix like R, there are S eigenvalues (i.e., $\lambda_1, \lambda_2, \ldots, \lambda_S$) to be solved.

Associated with each λ_i is an eigenvector, u_i; thus, there are S eigenvectors. The number of elements of u_i (a column vector) is equal to the order of R (the number of rows and columns). Thus, the dimensions of u_i are S rows by one column. The eigenvectors are obtained from the identity

$$Ru_i = \lambda_i u_i \qquad (19.6)$$

Solution of Eq. (19.6) is illustrated in Section 19.3. Although there are S eigenvalues and eigenvectors in R, in practice only the first two or three are usually of interest to the ecologist, since they represent the largest amount of variance explained and, it is hoped, produce an informative ordination. The percentage of total variance accounted for by each eigenvalue is computed by dividing each eigenvalue by the trace of the R matrix.

STEP 4. *SCALE EACH EIGENVECTOR.* The eigenvectors are usually scaled (or normalized) such that their *inner product* or sum of squares equals 1, that is,

$$u_i^t u_i = 1 \qquad (19.7a)$$

This is accomplished by computing a scaling factor, k, for each eigenvector. k_i is the reciprocal of the square root of the sum of squares of the elements of the ith eigenvector, that is,

$$k_i = \frac{1}{\sqrt{\sum_{q=1}^{S} u_{qi}^2}} \qquad (19.7b)$$

where u_{qi} is the element in the qth row of the ith eigenvector. Then, for each eigenvector

$$u_i = k_i \begin{bmatrix} u_{1,i} \\ u_{2,i} \\ \vdots \\ u_{s,i} \end{bmatrix} \qquad (19.7c)$$

The S normalized eigenvectors of R, each with S elements, are conveniently summarized as column vectors in the matrix $U_{S \times S}$.

STEP 5. *COMPUTE COORDINATES FOR A SPECIES ORDINATION.* The correlation of each species to the ith principal component is given by

$$v_i = u_i \sqrt{\lambda_i} \qquad (19.8)$$

These scaled correlations can also be used as coordinates for plotting the position of each species in a PCA ordination.

In matrix notation, the above relation can be written as

$$V_{S \times S} = U_{S \times S} \Lambda_{S \times S} \tag{19.9}$$

where the elements of V are the correlations of the ith species to the jth component and Λ is a matrix with λ_i's down the diagonal and 0's elsewhere.

STEP 6. *COMPUTE COORDINATES FOR AN SU ORDINATION.*   The coordinate positions of the SUs on the first three principal components are obtained by postmultiplying the transpose of A by its corresponding eigenvector. In matrix notation

$$Y_{N \times 3} = A_{N \times S}^t U_{S \times 3} \tag{19.10}$$

where the rows of Y are the coordinates for the N SUs on the first three principal components (element y_{ij} is the value for the ith SU on the jth principal component).

19.3   EXAMPLE: CALCULATIONS

The abundances of two species in five SUs will be used to illustrate the actual computations involved in PCA. These data, given in Table 19.1, constitute the community data matrix X. The use of such a simple data set makes it possible for us to readily demonstrate the numerous computations involved. Of course, we remind the student that only a large, multidimensional data set should be used in PCA. Although the computations illustrated below involve matrix algebra and are often tedious, the student is encouraged to proceed systematically with constant reference to Figure 19.1 and to Section 19.2; an appreciation of PCA will be the result.

STEP 1. *STANDARDIZATION OF THE DATA MATRIX.*   The community data matrix X is row centered using Eqs. (19.1) and (19.2). For species 1 with $\bar{x}_1 = 3$,

TABLE 19.1   *Data matrix X for the abundances of two species in five sampling units*

Species	SUs					Species	
	(1)	(2)	(3)	(4)	(5)	Means	Deviation[a]
(1)	2	5	5	3	0	3.0	18
(2)	0	3	4	2	1	2.0	10

[a]Sum of the squared deviations about the row mean.

TABLE 19.2 *The row (species) standardized matrix A: (a) by row centering and (b) by dividing by the scaling factor following Eqs. (19.1) and (19.2)*

(a)

Species	SUs (1)	(2)	(3)	(4)	(5)	Species Mean	Scaling Factor
$A = \begin{matrix}(1)\\(2)\end{matrix}$	-1 -2	$+2$ $+1$	$+2$ $+2$	0 0	-3 -1	0.0 0.0	4.24 3.16

(b)

$$A = \begin{bmatrix} -0.24 & +0.47 & +0.47 & 0.0 & -0.71 \\ -0.63 & +0.32 & +0.63 & 0.0 & -0.32 \end{bmatrix}$$

$$F_1 = \sqrt{(2-3)^2 + (5-3)^2 + \cdots + (0-3)^2} = 4.24$$

and, similarly, for species 2

$$F_2 = 3.16$$

The new, standardized elements of A are now computed by Eq. (19.1). For example, species 2 in SU 5 is

$$a_{2,5} = (1 - 2)/3.16 = -0.32$$

Each row element is now standardized by dividing its deviation from the row mean by the corresponding square root of the sum of squared deviations about the row mean. The complete A matrix is given in Table 19.2. Note that the row means are now zero.

The positions of the five SUs are shown in the original two-species coordinate space and in the species-centered space in Fig. 19.2a and 19.2b, respectively. Note that, in the species-centered case, SU 4 falls at the centroid, since its species abundances equal the row means.

STEP 2. COMPUTE SPECIES SIMILARITIES. Using Eq. (19.3), the correlations are:

$$R = \begin{bmatrix} -0.24 & +0.47 & +0.47 & 0.0 & -0.71 \\ -0.63 & +0.32 & +0.63 & 0.0 & -0.32 \end{bmatrix} \begin{bmatrix} -0.24 & -0.63 \\ +0.47 & +0.32 \\ +0.47 & +0.63 \\ 0.0 & 0.0 \\ -0.71 & -0.32 \end{bmatrix}$$

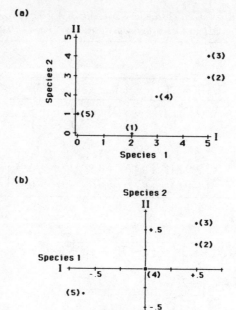

Figure 19.2   *The position of the five SUs (a) in two-species space and (b) in species-centered space.*

For row 1, column 1, that is, element $r_{1,1}$

$$r_{1,1} = (-0.24)(-0.24) + \cdots + (-0.71)(-0.71) = 1.0$$

and, similarly,

$$r_{1,2} = +0.82$$
$$r_{2,1} = +0.82$$
$$r_{2,2} = +1.0$$

(Note that $r_{1,2} = r_{2,1}$ since R is symmetric). In matrix form

$$R = AA^t = \begin{bmatrix} 1.0 & +0.82 \\ +0.82 & 1.0 \end{bmatrix}$$

STEP 3. COMPUTE EIGENVALUES AND EIGENVECTORS OF R.   From Eq. (19.5), the characteristic equation is given by

$$|R - \lambda I| = \left| \begin{bmatrix} 1.0 & +0.82 \\ +0.82 & 1.0 \end{bmatrix} - \lambda \begin{bmatrix} 1 & 0 \\ 0 & 1 \end{bmatrix} \right|$$

$$= \left| \begin{bmatrix} 1.0 & +0.82 \\ +0.82 & 1.0 \end{bmatrix} - \begin{bmatrix} \lambda & 0 \\ 0 & \lambda \end{bmatrix} \right|$$

$$= \left| \begin{bmatrix} 1 - \lambda & +0.82 \\ +0.82 & 1 - \lambda \end{bmatrix} \right| = 0$$

The determinant of a 2×2 matrix with elements a, b, c, and d is computed as

$$\left| \begin{bmatrix} a & b \\ c & d \end{bmatrix} \right| = (ad) - (bc)$$

Thus, the characteristic equation can be written as

$$(1 - \lambda)^2 - (0.82)^2 = 0$$

and, by rearranging, a second-order polynomial in λ is obtained

$$\lambda^2 - 2\lambda + 0.33 = 0$$

By using the quadratic equation, this polynomial equation is solved for the two eigenvalues as

$$\lambda = \frac{-2 \pm \sqrt{(-2)^2 - 4(0.33)}}{2} = \frac{-2 \pm 1.64}{2}$$

Thus,

$$\lambda_1 = \frac{-2 - 1.64}{2} = \frac{-3.64}{2} = 1.82 \text{ (ignore the sign)}$$

and

$$\lambda_{II} = \frac{-2 + 1.64}{2} = \frac{-0.36}{2} = 0.18 \text{ (ignore the sign)}$$

Note that the sum of the two eigenvalues equals the trace of R, which is the sum of the diagonal elements

$$\lambda_I + \lambda_{II} = r_{1,1} + r_{2,2} = 2$$

The percentage of the total variance of the observations accounted for by each of the eigenvalues is computed as a fraction of the trace:

$$\%\text{total variance } (\lambda_I) = \lambda_I \bigg/ \sum_{i=1}^{S} r_{i,i} = 1.82/2.0 = 91\%$$

$$\%\text{total variance } (\lambda_{II}) = \lambda_{II} \bigg/ \sum_{i=1}^{S} r_{i,i} = 0.18/2.0 = 9\%$$

Since all S eigenvalues were computed in this simple example, 100% of the variance is accounted for. In practice, this would not be the case.

The eigenvector associated with each eigenvalue is computed using Eq. (19.6). The eigenvector calculation for $\lambda_I = 1.82$ is

$$\begin{bmatrix} 1.0 & +0.82 \\ +0.82 & 1.0 \end{bmatrix} \begin{bmatrix} u_{1,I} \\ u_{2,I} \end{bmatrix} = 1.82 \begin{bmatrix} u_{1,I} \\ u_{2,I} \end{bmatrix}$$

From this, two equations in two unknowns are obtained:

$$1.0u_{1,I} + 0.82u_{2,I} = 1.82u_{1,I}$$

$$0.82u_{1,I} + 1.0u_{2,I} = 1.82u_{2,I}$$

and by rearranging

$$1.0u_{1,I} - 1.82u_{1,I} + 0.82u_{2,I} = 0$$

$$0.82u_{1,I} + 1.0u_{2,I} - 1.82u_{2,I} = 0$$

and

$$-0.82u_{1,I} + 0.82u_{2,I} = 0$$

$$+0.82u_{1,I} - 0.82u_{2,I} = 0$$

To solve these equations, we arbitrarily set $u_{1,I} = 1$ and then solve for $u_{2,I}$:

$$+0.82u_{2,I} = +0.82$$

$$-0.82u_{2,I} = -0.82$$

and, hence, $u_{2,1}$ is equal to 1. Thus, the eigenvector for eigenvalue λ_I is

$$u_I = \begin{bmatrix} 1.0 \\ 1.0 \end{bmatrix}$$

Similarly, the eigenvector associated with eigenvalue λ_{II} is

$$u_{II} = \begin{bmatrix} -1.0 \\ +1.0 \end{bmatrix}$$

STEP 4. SCALE EACH EIGENVECTOR. The scaling factor (k) for each u_i is, from Eq. (19.7b),

$$k_I = \frac{1}{\sqrt{1^2 + 1^2}} = 0.707$$

and

$$k_{II} = \frac{1}{\sqrt{-1^2 + 1^2}} = 0.707$$

For each u_i, from Eq. (19.7c),

$$u_I = 0.707 \begin{bmatrix} 1 \\ 1 \end{bmatrix} = \begin{bmatrix} 0.707 \\ 0.707 \end{bmatrix}$$

$$u_{II} = 0.707 \begin{bmatrix} -1 \\ 1 \end{bmatrix} = \begin{bmatrix} -0.707 \\ 0.707 \end{bmatrix}$$

Note, from Eq. (19.7),

$$u_I^t u_I = (0.707)(0.707) + (0.707)(0.707) = 1.0$$

and

$$u_{II}^t u_{II} = (-0.707)(-0.707) + (0.707)(0.707) = 1.0$$

The two eigenvectors are now conveniently written as column vectors in the U matrix:

$$U = \begin{bmatrix} 0.707 & -0.707 \\ 0.707 & 0.707 \end{bmatrix}$$

STEP 5. *COMPUTE COORDINATES FOR SPECIES ORDINATION.* From Eq. (19.8), the species correlations on principal components I and II are

$$v_I = \begin{bmatrix} 0.707 \\ 0.707 \end{bmatrix} \sqrt{1.82} = \begin{bmatrix} 0.95 \\ 0.95 \end{bmatrix}$$

$$v_{II} = \begin{bmatrix} -0.707 \\ 0.707 \end{bmatrix} \sqrt{0.18} = \begin{bmatrix} -0.30 \\ 0.30 \end{bmatrix}$$

Alternatively, from Eq. (19.9),

$$V = \begin{bmatrix} 0.707 & -0.707 \\ 0.707 & 0.707 \end{bmatrix} \begin{bmatrix} \sqrt{1.82} & 0 \\ 0 & \sqrt{1.82} \end{bmatrix} = \begin{bmatrix} 0.95 & -0.30 \\ 0.95 & 0.30 \end{bmatrix}$$

These correlations are summarized in Table 19.3. Both species have a high positive correlation with PCA axis I and a slight (and opposite) correlation with PCA axis II (Figure 19.3).

TABLE 19.3 *Eigenanalysis of the standardized A matrix giving (a) species coordinates (correlations), and (b) SU coordinates on the two principal components*

(a) Species Coordinates

Species	Principal Components	
	I	II
(1)	0.95	−0.30
(2)	0.95	0.30

(b) SU Coordinates

SU	Principal Components	
	I	II
(1)	−0.62	−0.28
(2)	0.55	−0.11
(3)	0.78	0.12
(4)	0.00	0.00
(5)	−0.72	0.28

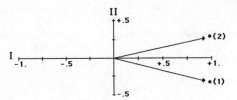

Figure 19.3 Species ordination showing the correlations of species 1 and 2 on principal components I and II.

STEP 6. COMPUTE COORDINATES FOR SU ORDINATION.   Using Eq. (19.10)

$$Y = A^t A = \begin{bmatrix} -0.24 & -0.63 \\ 0.47 & 0.32 \\ 0.47 & 0.63 \\ 0.00 & 0.00 \\ -0.71 & -0.32 \end{bmatrix} \begin{bmatrix} 0.707 & -0.707 \\ 0.707 & 0.707 \end{bmatrix}$$

where the SU coordinates for principal component I (column 1) are

$$y_1 = (-0.24)(0.707) + (-0.63)(0.707) = -0.62$$
$$\vdots$$
$$y_5 = (-0.71)(0.707) + (-0.32)(0.707) = -0.72$$

and the SU coordinates for principal component II (column 2) are

$$y_1 = (-0.24)(-0.707) + (-0.63)(0.707) = -0.28$$
$$\vdots$$
$$y_5 = (-0.71)(-0.707) + (-0.32)(0.707) = 0.28$$

These SU coordinates are summarized in Table 19.3*b* and the resultant SU ordination is shown in Figure 19.4. Compare this with the original coordinate system (Figure 19.2) and note how the two-species space (two-axes system) can be represented by a single axis (principal component I) with very little distortion (accounts for 91% of the variation) in the relationships between the SUs.

It is obvious from the tedious calculations involved for even a very simple data set (two species and five SUs) that computerized algorithms are essential

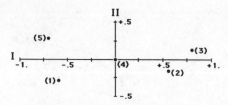

Figure 19.4 *Position of five SUs on principal components I and II.*

for a PCA. A number of numerical methods are available for extracting eigenvalues and eigenvectors from a square symmetrical *R*-mode or *Q*-mode matrix. Statistical packages, such as SAS (Ray 1982) and BMDP (Dixon and Brown 1979), include PCA and are easy to use. However, the student is cautioned to first use a simple data set of known properties (such as the one

TABLE 19.4 *Principal components analysis for six Panamanian localities (SUs) using abundance data for five cockroach species*

R Matrix (Species Correlations—not shown)

First three eigenvalues of the *R* matrix = 3.026   1.175   0.790
Eigenvectors:

Vector 1	0.154	−0.255	−0.564	−0.564	−0.543
Vector 2	0.755	−0.589	0.100	0.192	0.193
Vector 3	−0.572	−0.708	0.173	0.260	−0.271

Percentage of trace by each eigenvalue: 60.5%   23.5%   15.8%
Accumulated percentage by the eigenvalues: 60.5%   84.0%   99.8%

SU coordinates on the first three components

SU	I	II	III
(1)	−1.508	0.206	0.187
(2)	−0.175	−0.660	−0.593
(3)	0.454	−0.112	0.314
(4)	0.477	−0.116	0.307
(5)	0.326	−0.128	0.198
(6)	0.425	0.809	−0.413

Species correlations on the first three components

Species	I	II	III
(1)	0.268	0.818	−0.509
(2)	−0.443	−0.638	−0.629
(3)	−0.981	0.108	0.154
(4)	−0.950	0.208	0.231
(5)	−0.945	0.209	−0.241

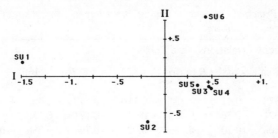

Figure 19.5   *Principal components ordination of six Panamanian localities (SUs).*

used here) to become familiar with the output produced with these packages before attempting PCA on a complex set of data. These "canned" programs often generate more output than needed for most ecological applications. However, on the positive side, these programs usually provide options for graphical output, which is essential for examining the similarity relationships among the SUs.

19.4   EXAMPLE: PANAMANIAN COCKROACHES

Using the Panamanian cockroach data matrix of five species abundances in six SUs (Table 11.4a) a R-strategy PCA ordination was performed with the BASIC computer program PCA.BAS (see accompanying disk). The results are given in Table 19.4.

The SU coordinates can be used to display graphically the six localities within a coordinate system where the relative positions of the localities reflect similarities (Figure 19.5). Principal component I accounts for over 60% of the variation. Localities 3, 4, and 5 are very similar; all are characterized by a low abundance of all five species. Locality 6 contains a single species (1), while locality 1 has high abundances of species 3, 4, and 5. For simplicity of Figure 19.5, species are not shown and only two components are graphed. Further interpretation of this ordination is deferred to Chapter 24.

19.5   EXAMPLE: WISCONSIN FORESTS

An R-strategy PCA ordination using PCA.BAS was performed on the Wisconsin forest data of eight tree abundances in 10 sites (Table 11.6a). The results of the SU ordination are given in Table 19.5 and Figure 19.6. For clarity, the species correlations (positions) with axes are not shown, but

TABLE 19.5 *Principal components analysis for 10 upland forest sites (SUs), based on the abundances of eight trees*

R Matrix (Species Correlations—see Table 12.2)

First three eigenvalues of the R matrix = 4.472   1.373   1.166

Eigenvectors:

Vector 1	0.378	0.415	0.301	−0.116	0.119	−0.461	−0.387	−0.453
Vector 2	−0.021	0.364	−0.409	−0.775	−0.171	0.110	0.186	0.151
Vector 3	−0.308	0.101	0.313	0.021	−0.794	0.080	−0.389	0.091

Percentage of trace by each eigenvalue: 55.9%   17.2%   14.6%
Accumulated percentage by the eigenvalues: 55.9%   73.1%   87.6%

SU coordinates on the first three components

SU	I	II	III
(1)	0.743	0.310	0.051
(2)	0.691	0.293	0.050
(3)	0.744	0.369	0.136
(4)	0.715	−0.263	−0.068
(5)	0.330	−0.710	−0.338
(6)	−0.413	−0.438	0.648
(7)	−0.452	−0.196	−0.535
(8)	−0.665	−0.065	0.456
(9)	−0.942	0.386	−0.077
(10)	−0.751	0.313	−0.322

Species correlations on the first three components

Species	I	II	III
(1)	0.798	−0.025	−0.333
(2)	0.877	0.427	0.109
(3)	0.636	−0.479	0.338
(4)	−0.246	−0.908	−0.023
(5)	0.253	−0.200	−0.857
(6)	−0.975	−0.129	0.086
(7)	−0.818	0.218	−0.420
(8)	−0.958	0.178	0.098

are given in Table 19.5, indicating that the oaks are positively correlated with principal component I and basswood, ironwood, and maple are negatively correlated with principal component I. The SUs are clearly separated on component I based on their abundances of these species. SUs 1–4 are dominated by oaks, and SUs 8–10 are dominated by basswood, ironwood, and maple. SUs 5–7 have more of a mixture of species. Black oak and red oak are positively and negatively correlated with component II, respectively. Thus,

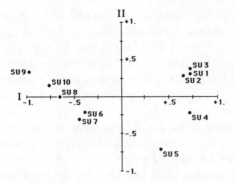

Figure 19.6 *Principal components ordination of 10 upland forest sites (SUs) in southern Wisconsin.*

SU 5, with its abundance of red oak, is moved in the negative direction on II, while SUs 1–3, with abundant black oak, are moved in the positive direction on component II. SU 4 has some black oak, but abundant red oak moves it in the negative direction on component II. Further interpretation of this ordination is provided in Chapter 24.

19.6   ADDITIONAL TOPICS IN PRINCIPAL COMPONENT ORDINATION

We illustrated PCA by following an *R*-strategy where the desired SU ordination was derived from a matrix of species correlations rather than from a matrix of SU similarities (the *Q*-strategy). In most ecological applications this will be the most efficient strategy, since the rank of *R* (number of species) will typically be less than the rank of *Q* (number of SUs). However, if the number of species exceeds the number of SUs, the *Q*-strategy can be followed (Figure 19.1). The resultant SU ordination will be the same as long as the *Q* matrix is derived from a species-centered (row-centered) *A* matrix as described in Section 19.2.

It is a mistake to derive the *Q* matrix from a SU-centered (column-centered) *A* matrix by transposing the original data matrix and substituting it into a "canned" program within a statistical package. In so doing, the eigenvalues and vectors of the PCA are computed on a SU (rather than a species) correlation matrix, and the resultant SU ordination will *not* be the same as that from the recommended strategy described in this chapter. In fact, the first axis will *not* coincide with the maximal variation, resulting in a loss of ordination efficiency (Orloci 1966).

An alternative to row centering is to perform the PCA calculations on a *noncentered* data matrix (Carleton 1980, Noy-Meir 1973, Noy-Meir et al. 1975). Some useful ecological information may be obtained by using a non-centering ordination. The first principal component extracts variation in the data due to overall species-abundance patterns, which are implicitly removed by species centering. These abundance patterns may be of considerable interest in a community study.

Another alternative is to *double transform* the data matrix before the PCA calculations. Double transformation involves dividing each element (observation) in the data matrix by the square root of its row total and by the square root of its column total (see Chapter 20 for details), leading to a *correspondence analysis* through eigenanalysis calculations (Hill 1973c, 1974).

In Eq. (19.2) a scaling factor F_i was given for computing the elements of the A matrix. Goff (1975) conducted a study to compare the effect of various values for F_i on R-strategy ordinations. With $F_i = 1$, the elements of the R matrix are dispersion coefficients. With $F_i = \sqrt{N-1}$, the elements of the R matrix are variances and covariances. A PCA on a dispersion or variance–covariance matrix will give results similar to that for a correlation matrix, differing only in scale, since the only difference is the value of the divisor F_i. While the correlation coefficient was found to give relatively greater significance to rarer species, the overall results were similar to dispersion and covariance, at least for Goff's data. This again points out the potential importance of transformations (implicit in many resemblance measures) on an analysis and the subsequent ecological interpretations.

In both the R- and Q-strategies presented in this chapter, the resemblance measures between species or SUs were derived as function of scalar products [Eq. (19.1)–(19.4)]. Gower (1966) showed that it is possible to expand this to include Euclidean distances between SUs. Orloci (1973) terms this a *D-strategy PCA*. The distance matrix D has elements given by

$$D_{N,N} = -0.5[\text{SED}_{j,k}] \tag{19.11}$$

where $\text{SED}_{j,k}$ is the squared Euclidean distance between SUs j and k (see Chapter 14, Section 14.2). The eigenvalues and eigenvectors of D are computed, thus obtaining an SU ordination. Thus, it is possible to perform a PCA on R, Q, or D matrices.

Extensions of the D-strategy, called *principal coordinates analysis* (PCO), allows eigenanalysis to be used to reduce the dimensionality of the D matrix to give SU relationships in a Euclidean space. The details of PCO are beyond the scope of this text, but the interested student is referred to the original paper by Gower (1966) and the comprehensive treatments by Legendre and Legendre (1983), Orloci (1978), and Pielou (1984).

19.7 SUMMARY AND RECOMMENDATIONS

1. Principal components analysis (PCA) has appeal as an ecological ordination method because it is a multivariate eigenanalysis method that produces an SU ordination by extracting axes of maximum variation from a *R*-mode (species) resemblance matrix or a *Q*-mode (SU) resemblance matrix (Section 19.1). Also, PCA is readily available, being included in most statistical computer packages.

2. Of the possible strategies for a PCA to produce an SU ordination, we recommend Orloci's strategy where the community data are species centered, normalized, and converted to an interspecific correlation matrix before eigenanalysis (Section 19.2). This strategy is conceptually straightforward and computationally efficient.

3. Although the mathematical manipulations for an eigenanalysis on even a very simple community data matrix are tedious and require some understanding of matrix algebra, we strongly recommend that the student work through the example computations (Section 19.3) in order to gain a basic understanding of PCA before using canned computer programs.

4. The example computations (Section 19.3) and the cockroach and forest examples (Sections 19.4 and 19.5, respectively), illustrate how PCA "rotates" the original data "species space" into a new space of reduced dimensionality (i.e., a two- or three-axis ordination) to expose underlying community patterns. It is hoped that these patterns will have an ecological interpretation and will lead to testable hypotheses about community structure (see Chapter 24).

5. If there are strong nonlinearities in the data, PCA (being a linear model) will poorly represent the true SU relationships. Nonlinear ordination models are described in Chapter 21.

6. When planning to use PCA ordination, we recommend that the basic PCA procedure described in this chapter be used in conjunction with other PCA strategies, for example, eigenanalysis on a noncentered data matrix or on an SU distance matrix (Section 19.6). Results of different approaches can be compared for consistency in the community patterns revealed. If large differences emerge, then one can investigate whether these differences are due to the methods or are inherent in the data.

CHAPTER 20

Correspondence Analysis

Correspondence analysis (COA), and its variant, detrended correspondence analysis (DCA), are now widely used ordination techniques. One of the strong attractions of COA is that "corresponding" SU and species ordinations are obtained simultaneously, allowing the community ecologist to examine the ecological interrelationships between SUs and species in a single analysis. COA can be derived through the use of an eigenanalysis approach (similar to PCA) or through a series of weighted-average operations. (COA is often called *reciprocal averaging* when this latter procedure is used.) In this chapter we describe COA using the eigenanalysis approach introduced in the previous chapter. In the next chapter we briefly describe DCA as one ordination approach to the nonlinear "problem."

20.1 GENERAL APPROACH

In a study of forest communities of southern Wisconsin, Curtis and McIntosh (1951) conducted one of the first examples of what we referred to in Chapter 17 as an *indirect ordination*. They arbitrarily assigned "weights" (called *climax adaptation numbers*) to forest species based on whether the species were characteristic of early to late successional seres. These weights ranged from 1 (for an early successional species like bur oak) to 10 (for a climax species like sugar maple). Based on the species composition of a particular stand [sampling unit (SU)], a score was obtained as a weighted average of these weights times species abundances. Using the scores for all stands, a one-dimensional SU ordination, reflecting a succession gradient, was obtained. This type of ordination is referred to as *weighted average ordination* (Whittaker 1967).

Conceptually, an ordination of SUs into a reduced *species space* (Figure 13.2) simply involves computing *weightings* for each species and then obtaining a *score* for each SU based on the sum of the species weightings times the species abundances; hence, the SU score or coordinate is a weighted sum. This is done for each dimension of interest in the reduced space. Basically, different ordination procedures differ only in how they obtain the species weights and the SU scores (Pielou 1984). For example, recall from Chapter 19 that PCA coordinates (scores) for SUs on the ordination axes (principal components) were derived by extracting eigenvalues and vectors from a species correlation matrix. The eigenvectors contained the weights for each species for use in computing the SU scores.

When using the eigenanalysis approach, COA can be viewed as simply a variant of PCA (Pielou 1984). However, COA differs from PCA in two important ways: (1) the way the original data are transformed and (2) the way the eigenvectors are computed into SU coordinates or scores. In addition to the SU coordinates, the corresponding species coordinates or correlations are also obtained in a COA, thus providing a simultaneous SU and species ordination. The species ordination coordinates are averages of the SU ordination scores and, vice versa, the SU ordination coordinates are averages of the species ordination correlations (Gauch 1982).

20.2 PROCEDURES

The strategy for calculating SU and species scores using COA is shown in Figure 20.1. Following Chapter 19, the same matrix symbols of Orloci (1966,

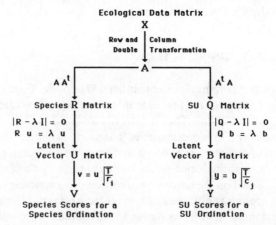

Figure 20.1 *Strategy for a correspondence analysis to obtain a simultaneous SU and species ordination.*

1967b) are used: the ecological data matrix X has S species (rows) and N SUs (columns); the species-by-species R matrix has a rank of S and the SU-by-SU Q matrix has a rank of N; the ith and jth elements of the X matrix are denoted as x_{ij}; the elements of the A matrix are a_{ij}; and so forth. The student is referred to Chapter 19 for a detailed description of the eigenanalysis procedure.

STEP 1. TRANSFORMATION OF THE DATA MATRIX X. First, sum the abundances for each of the $i = 1$ to S species across the N SUs (i.e., the row totals):

$$r_i = \sum_{j=1}^{N} x_{ij} \tag{20.1}$$

Second, sum the abundances for each of the $j = 1$ to N SUs down the S species (i.e., the column totals):

$$c_j = \sum_{i=1}^{S} x_{ij} \tag{20.2}$$

Third, calculate the total abundance by summing either the row totals or the column totals:

$$T = \sum_{i=1}^{S} r_i = \sum_{j=1}^{N} c_j \tag{20.3}$$

This total abundance, T, will be used later for transforming the eigenvectors into SU and species scores on COA components. Lastly, divide the abundances of the ith species in the jth SU by the square root of the ith row total (r_i) and the square root of the jth column total (c_j):

$$a_{ij} = \frac{x_{ij}}{\sqrt{r_i}\sqrt{c_j}} \tag{20.4}$$

Thus, by using Eq. (20.4), the matrix $A_{S \times N}$ is formed by a double transformation of the data matrix $X_{S \times N}$ (see Figure 20.1).

STEP 2. COMPUTE SPECIES (R) AND SU (Q) RESEMBLANCE MATRICES. This step (see Figure 20.1) is identical to Step 2 for a PCA; the student is referred to Section 19.2 for details. Be aware, however, that although similar in character, the R and Q matrices generated here from the A matrix are different from those in PCA because of the different transformation used to generate A.

STEP 3. COMPUTE EIGENVALUES AND EIGENVECTORS OF R AND Q. Again, the procedure for computing eigenvalues and vectors of square matrices are described in Step 3, Section 19.2 of the PCA chapter. A unique feature of the Q and R matrices generated from double transformation is that the eigenvalues are identical (Orloci 1978, Pielou 1984).

STEP 4. COMPUTE SPECIES AND SU ORDINATIONS. The species correlations or coordinates, v, on the COA components are calculated by scaling the eigenvectors, u, as

$$v_i = u_i \sqrt{\frac{T}{r_i}} \tag{20.5}$$

The SU scores or coordinates, y, on the COA components are calculated by scaling the eigenvectors, b, as

$$y_j = b_j \sqrt{\frac{T}{c_j}} \tag{20.6}$$

The scaled species vectors, v, are summarized as column vectors in the V matrix and the scaled SU vectors, y, are summarized as column vectors in the Y matrix (see Figure 20.1). The results can be plotted for ease of interpretation.

20.3 EXAMPLE: CALCULATIONS

The computations involved in a complete COA will be illustrated here by using the ecological data matrix for the abundances of three species in five SUs (Table 20.1a). We say "complete COA" because, following the strategy shown in Figure 20.1, one can choose to calculate only SU scores for an SU ordination and not do the "corresponding" calculations for a species ordination. Here the calculations for both SU and species coordinates are illustrated. However, details for the rather lengthy and tedious eigenvalue and eigenvector calculations are not provided, as this would be repetitive of Chapter 19.

STEP 1. TRANSFORMATION OF THE DATA MATRIX X. First obtain the required species (row) and SU (column) sums [Eqs. (20.1) and (20.2)] along with the grand total [Eq. (20.3)] for the data matrix (Table 20.1a). Then each species abundance value is transformed as in Eq. (20.4). For example, the transformed abundance value for species 1 in SU 1 would be

$$a_{1,1} = 2/(\sqrt{15}\sqrt{4}) = 0.258$$

The row and column (double) transformed matrix A is shown in Table 20.1b.

TABLE 20.1 *(a) The community data matrix X for the abundances of three species in five sampling units (SUs), along with row and column totals and the grand total and (b) the transformed community data matrix A derived from X by dividing the abundances of three species in five sampling units (SUs) by the square root of row and column totals*

(a)

			SUs			
Species	(1)	(2)	(3)	(4)	(5)	Row Sum
(1)	2	5	5	3	0	15
X = (2)	0	3	4	2	1	10
(3)	2	0	1	0	2	5
Column sum	4	8	10	5	3	T = 30

(b)

		SUs			
Species	(1)	(2)	(3)	(4)	(5)
(1)	0.258	0.456	0.408	0.346	0.000
A = (2)	0.000	0.335	0.400	0.283	0.183
(3)	0.447	0.000	0.141	0.000	0.516

STEP 2. COMPUTE SPECIES (R) AND SU (Q) RESEMBLANCE MATRICES. The species scalar-product R matrix is formed by postmultiplying A by its transpose and the SU scalar-product Q matrix is obtained by premultiplying A by its transpose (see Figure 20.1). The R and Q matrices are given in Table 20.2.

STEP 3. COMPUTE EIGENVALUES AND EIGENVECTORS OF R AND Q. The three eigenvalues of R are

$$\lambda_I = 1.00, \quad \lambda_{II} = 0.384, \quad \text{and} \quad \lambda_{III} = 0.050$$

and the eigenvalues of Q are

$$\lambda_I = 1.00, \quad \lambda_{II} = 0.384, \quad \text{and} \quad \lambda_{III} = 0.050$$

which are equal to those for R, as expected.

The eigenvectors associated with each of eigenvalues of R and of Q are given as column vectors in the matrices U and B, respectively (Table 20.3).

STEP 4. COMPUTE SPECIES AND SU ORDINATIONS. The species coordinates (correlations) on the COA components are calculated by multiplying each column vector of U by $\sqrt{T/r_i}$ [Eq. (20.5)]. For example, the correlation for species 3 on component III is

TABLE 20.2 *Scalar-product similarity matrices for (a) species (R-mode) and (b) for SUs (Q-mode) based on the A matrix (see Table 20.1b)*

	(a) For Species (R-mode)		
Species	(1)	(2)	(3)
(1)	0.56	0.41	0.17
R = (2)	0.41	0.39	0.15
(3)	0.17	0.15	0.49

	(b) For SUs (Q-mode)				
SUs	(1)	(2)	(3)	(4)	(5)
(1)	0.27	0.12	0.17	0.09	0.23
(2)	0.12	0.32	0.32	0.25	0.06
Q = (3)	0.17	0.32	0.35	0.25	0.15
(4)	0.09	0.25	0.25	0.20	0.05
(5)	0.23	0.06	0.15	0.05	0.30

TABLE 20.3 *Matrices of eigenvectors for eigenvalues associated with (a) the R matrix and (b) the Q matrix (see Table 20.2)*

	(a) The R Matrix		
	Eigenvectors		
Species	I	II	III
(1)	0.707	−0.331	0.625
U = (2)	0.577	−0.240	−0.781
(3)	0.408	−0.913	0.022

	(b) The Q Matrix		
	Eigenvectors		
SUs	I	II	III
(1)	0.365	−0.520	−0.765
(2)	0.516	0.374	−0.104
B = (3)	0.577	0.165	0.242
(4)	0.408	0.294	0.020
(5)	0.316	−0.690	0.587

TABLE 20.4 *Matrices of (a) species correlations, V, and (b) SU coordinates, Y, on three COA components derived from the U and B matrices, respectively (see Table 20.3).*

(a) Species Correlations, V

	COA Components		
Species	I	II	III
(1)	1.00	−0.469	0.883
V = (2)	1.00	−0.415	−1.35
(3)	1.00	2.24	0.054

(b) SU Coordinates, Y

	COA Components		
SUs	I	II	III
(1)	1.00	−1.42	−2.10
(2)	1.00	0.723	−0.202
Y = (3)	1.00	0.285	0.419
(4)	1.00	0.721	0.048
(5)	1.00	−2.18	1.86

$$v_{3,3} = 0.022\sqrt{30/5} = 0.054$$

The SU coordinates on the COA components are calculated by multiplying each column vector of B by $\sqrt{T/c_j}$. For example, the coordinate for SU 5 on component II is

$$y_{5,2} = -0.690\sqrt{30/3} = -2.18$$

The species and SU coordinates on the first three COA components are given in Table 20.4. Note that the species correlations on the first COA component are always 1.0, a consequence of the double transformation and subsequent rescaling used. This same property arises when using an alternative method of computing COA, known as *reciprocal averaging* (Legendre and Legendre 1983, Pielou 1984). Thus, the first COA component is ignored when constructing the species ordination. As in the case of the species coordinates, the SU coordinates on the first COA component are equal to 1.0 and are ignored for the SU ordination.

The COA ordination of SUs (Figure 20.2) is very similar to that for the PCA ordination (Figure 19.4), the differences being largely because of the use of only two species with the PCA example.

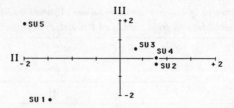

Figure 20.2   *Correspondence analysis ordination of five SUs on components II and III.*

TABLE 20.5   *Results for a correspondence analysis on (a) six Panamanian localities (SUs) and (b) five cockroach species*

(a)   Scalar-product (Q-mode) similarities between SUs

SUs	(1)	(2)	(3)	(4)	(5)	(6)
(1)	0.58	0.30	0.07	0.02	0.17	0.23
(2)	0.30	0.45	0.06	0.08	0.15	0.12
$Q =$ (3)	0.07	0.06	0.02	0.01	0.02	0.00
(4)	0.02	0.08	0.01	0.02	0.02	0.00
(5)	0.17	0.15	0.02	0.02	0.08	0.00
(6)	0.23	0.12	0.00	0.00	0.00	0.22

Eigenvalues for Q (first three)

$$\lambda_I = 1.000 \qquad \lambda_{II} = 0.239 \qquad \lambda_{III} = 0.115$$

Eigenvectors for Q (first three)

SUs	(1)	(2)	(3)	(4)	(5)	(6)
Vectors 1:	0.710	0.553	0.092	0.065	0.263	0.328
2:	−0.470	0.764	0.093	0.228	0.029	−0.366
3:	0.492	−0.111	0.260	0.022	−0.191	−0.801

SU coordinates on the first three components

SUs:	(1)	(2)	(3)	(4)	(5)	(6)
Components I:	1.00	1.00	1.00	1.00	1.00	1.00
II:	−0.66	1.38	1.00	3.47	0.11	−1.11
III:	0.69	−0.20	2.80	0.34	−0.73	−2.44

TABLE 20.5 (*continued*)

(b)　　　　　Scalar-product (*R-mode*) similarities between species

Species	(1)	(2)	(3)	(4)	(5)
(1)	0.04	0.00	0.00	0.00	0.08
(2)	0.41	0.42	0.14	0.04	0.30
R = (3)	0.17	0.15	0.22	0.11	0.28
(4)	0.00	0.04	0.11	0.06	0.13
(5)	0.08	0.30	0.28	0.13	0.62

Eigenvalues for *R* (first three)

$$\lambda_I = 1.000 \qquad \lambda_{II} = 0.239 \qquad \lambda_{III} = 0.115$$

Eigenvectors for *R* (first three)

Species	(1)	(2)	(3)	(4)	(5)
Vectors 1:	0.065	0.500	0.383	0.174	0.754
2:	−0.150	0.849	−0.274	−0.235	−0.356
3:	−0.471	0.057	0.684	0.354	−0.426

Species coordinates on the first three COA components

Species	(1)	(2)	(3)	(4)	(5)
Components I:	1.00	1.00	1.00	1.00	1.00
II:	−2.29	1.70	−0.72	−1.36	−0.47
III:	−7.18	0.12	1.79	2.04	−0.56

20.4 EXAMPLE: PANAMANIAN COCKROACHES

Using the Panamanian cockroach data matrix (Table 11.4*a*) and the BASIC program COA.BAS (on accompanying disk), a COA ordination of SUs and species was performed. The results are given in Table 20.5 and the SU ordination is shown in Figure 20.3. For these data, the patterns of the SUs in the COA are only somewhat comparable to those for PCA (see Figure 19.5) and we exercise the option of leaving it to the student to try to interpret the differences!

20.5 EXAMPLE: WISCONSIN FORESTS

The results of a COA for the Wisconsin upland forest data (Table 11.6*a*) are given in Table 20.6. These results were obtained with the BASIC program COA.BAS. The positions of the SUs in the COA ordination (Figure 20.4) are

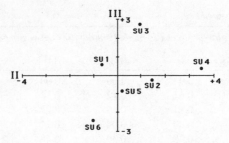

Figure 20.3 *Correspondence analysis ordination of the six Panamanian cockroach localities (SUs).*

TABLE 20.6 *Results for a correspondence analysis on (a) 10 upland forest sites (SUs) and (b) eight tree species*

(a) Scalar-product (Q-mode) similarities between SUs

SUs	(1)	(2)	(3)	(4)	(5)	(6)	(7)	(8)	(9)	(10)
(1)	0.19	0.18	0.15	0.15	0.11	0.04	0.07	0.03	0.01	0.02
(2)	0.18	0.19	0.15	0.16	0.10	0.04	0.07	0.03	0.02	0.02
(3)	0.15	0.15	0.18	0.16	0.09	0.05	0.06	0.04	0.01	0.04
(4)	0.15	0.16	0.16	0.16	0.13	0.07	0.09	0.06	0.03	0.05
(5)	0.11	0.10	0.09	0.13	0.16	0.10	0.12	0.07	0.05	0.07
(6)	0.04	0.04	0.05	0.07	0.10	0.16	0.11	0.15	0.12	0.12
(7)	0.07	0.07	0.06	0.09	0.12	0.11	0.17	0.14	0.16	0.15
(8)	0.03	0.03	0.04	0.06	0.07	0.15	0.14	0.17	0.16	0.16
(9)	0.01	0.02	0.01	0.03	0.05	0.12	0.16	0.16	0.20	0.18
(10)	0.02	0.02	0.04	0.05	0.07	0.12	0.15	0.16	0.18	0.19

Eigenvalues for Q (first three)

$$\lambda_I = 1.00 \qquad \lambda_{II} = 0.54 \qquad \lambda_{III} = 0.096$$

Eigenvectors for Q (first three)

SUs:	(1)	(2)	(3)	(4)	(5)	(6)	(7)	(8)	(9)	(10)
1:	0.30	0.30	0.29	0.33	0.32	0.30	0.36	0.32	0.30	0.32
2:	0.39	0.39	0.35	0.30	0.11	−0.20	−0.18	−0.31	−0.42	−0.36
3:	0.21	0.24	0.18	−0.14	−0.58	−0.54	0.08	−0.06	0.35	0.29

SU coordinates on the first three components

SUs:	(1)	(2)	(3)	(4)	(5)	(6)	(7)	(8)	(9)	(10)
I:	1.00	1.00	1.00	1.00	1.00	1.00	1.00	1.00	1.00	1.00
II:	1.28	1.27	1.22	0.91	0.35	−0.66	−0.51	−0.98	−1.38	−1.11
III:	0.69	0.80	0.64	−0.41	−1.82	−1.78	0.21	−0.20	1.15	0.90

TABLE 20.6 *(continued)*

(b)	Scalar-product (R-mode) similarities between species							
Species	(1)	(2)	(3)	(4)	(5)	(6)	(7)	(8)
(1)	0.23	0.22	0.17	0.14	0.13	0.03	0.03	0.02
(2)	0.22	0.30	0.19	0.09	0.12	0.00	0.00	0.00
(3)	0.17	0.19	0.24	0.18	0.14	0.11	0.05	0.09
(4)	0.14	0.09	0.18	0.21	0.13	0.16	0.10	0.14
(5)	0.13	0.12	0.14	0.13	0.15	0.08	0.08	0.08
(6)	0.03	0.00	0.11	0.16	0.08	0.21	0.16	0.22
(7)	0.03	0.00	0.05	0.10	0.08	0.16	0.18	0.19
(8)	0.02	0.00	0.09	0.14	0.08	0.22	0.19	0.25

Eigenvalues for R (first three)

$$\lambda_I = 1.00 \qquad \lambda_{II} = 0.54 \qquad \lambda_{III} = 0.096$$

Eigenvectors for R (first three)

Species:	(1)	(2)	(3)	(4)	(5)	(6)	(7)	(8)
1:	0.35	0.33	0.43	0.41	0.33	0.34	0.27	0.34
2:	0.39	0.53	0.20	−0.07	0.08	−0.40	−0.36	−0.47
3:	0.08	0.50	−0.39	−0.54	0.02	−0.08	0.48	0.25

Species coordinates on the first three components

Species:	(1)	(2)	(3)	(4)	(5)	(6)	(7)	(8)
I:	1.00	1.00	1.00	1.00	1.00	1.00	1.00	1.00
II:	1.11	1.61	0.47	−0.17	0.25	−1.17	−1.32	−1.37
III:	0.24	1.51	−0.91	−1.30	0.05	−0.24	1.77	0.73

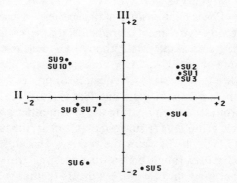

Figure 20.4 *Correspondence analysis ordination of 10 forest sites (SUs) in southern Wisconsin.*

very similar to those from a PCA on these data (Figure 19.6). The major
differences in the COA is that of separating SUs 9 and 10 from 8. From
Table 11.6a, it can be seen that SUs 8, 9, and 10 all have a high abundance
of basswood, ironwood, and maple (species 6, 7, and 8). However, SU 8 has
a strong oak influence (species 3 and 4).

COA also separates SU 6 from 7 and, again from Table 11.6a, it can be
seen that SU 6 has a dominance of species 3, 4, 5, 6, and 8, whereas SU 7 also
has these species and, in addition, species 1 and 7. Thus, SU 7 is weighted
differently in the COA ordination. Overall, the COA seems to give a sharper
delimitation of relationships between SUs than PCA, at least for these data.

20.6 ADDITIONAL TOPICS IN CORRESPONDENCE ANALYSIS

COA uses a *double transformation* as an alternative to the *centered-normalized
transformation* used in PCA, and it also scales the eigenvectors differently.
A main feature of COA is that coordinates are obtained for both SUs and
species so that "corresponding" SU and species ordinations can be constructed.
The SU and species coordinates are obtained in such a way as to maximize
their intercorrelation. COA ordination should optimally fit ordination theory,
that is, position SUs and species within a coordinate system to best reflect
their similarity patterns. The ecological effectiveness of COA as an ordination
method is clearly shown in examples provided by Gauch (1982) and COA has
become a popular ordination method (see Table 17.1).

In reviewing COA in relation to polar ordination (PO), Beals (1984) con-
cludes that COA "is good primarily for only one axis," documenting from
the literature that "the second axis tends to be an arch, and axes after the first
are often hard to interpret." As illustrated in the next chapter, the arch
problem is a distinctly curved pattern of SUs within a PCA or COA ordination
resulting from nonlinear relationships of the species—usually observed when
sampling over broad environmental gradients, that is, community data with
high beta diversity (see Chapter 8). Although less sensitive to nonlinearities
in community data than PCA, COA nonetheless has the "arch" problem
(Gauch 1982). The effectiveness of COA can be improved by *detrending* when
dealing with nonlinear data sets (Hill and Gauch 1980; see Chapter 21).

Conclusions from studies comparing COA with other ordination methods
are mixed. Gauch et al. (1977) suggest that PO is superior to COA, yet,
Whittaker and Gauch (1973) claim the reverse. This difference appears to
be related to COA performing better when dealing with community data
with one dominant gradient of high beta diversity, that is, covering a broad
environmental gradient (Beals 1984). Beals also suggests that arguments
claiming COA is better because SUs are equitably spread along the first axis

is actually a weakness because this attribute may obscure real discontinuities in the gradient underlying the first axis. We support Beals's conclusion that there is no one "unequivocally" best ordination method, and we recommend, as before, that the student use more than one ordination method on a given set of data and examine results for consistency.

20.7 SUMMARY AND RECOMMENDATIONS

1. Correspondence analysis (COA) is a weighted-average technique that reciprocally double-transforms community data (by species and SUs) and then employs eigenanalysis to produce "corresponding" species and SU ordinations (Section 20.1).

2. The procedures and computations for COA (Section 20.2 and 20.3, respectively) are very similar to those used in principal components analysis (PCA); we recommend the student carefully work through these in order to gain a basic understanding of COA before using "canned" computer programs.

3. The examples (Sections 20.4 and 20.5) illustrate that in some cases COA produces ordination patterns very similar to those for PCA (Wisconsin forest data), but in other cases, COA gives quite different results (Panamanian cockroach data).

4. Comparative studies suggest that COA performs better than PO and PCA when there is one dominant and relatively broad underlying gradient. The second axis often has the "arch" problem caused by nonlinear species responses along the gradient (we refer the student to the next chapter) and is often difficult to interpret ecologically.

5. Because COA, PCA, and PO methods differ in the way they "tackle" community data (but not in their aim), we recommend that these, and perhaps other, ordination methods be applied to the same data and any differences be used to gain insights into existing community patterns.

CHAPTER 21

Nonlinear Ordinations

Community ecologists often collect data over wide environmental gradients where sampling units (SUs) occurring at opposite ends of these gradients have few or no species in common. It can be readily demonstrated that over such gradients individual species-abundance patterns will typically be *nonlinear* (e.g., bell-shaped) and that these nonlinear abundances can cause the distribution of SUs in species hyperspace to be "arched" (or even spiraled). This nonlinear/arch problem is a serious limitation to the methods described in the previous ordination chapters. There are two solutions to this problem: (1) use an ordination method that removes (detrends) the SU arch in cases where *moderate* species nonlinearities cause a distinct arch or horseshoe pattern, or (2) use an ordination method that orders (ranks, catenates) the SU spiral in cases where *strong* species nonlinearities cause spiral or looping patterns. In this chapter we describe polynomial ordination, which detrends, and nonmetric multidimensional scaling, which rank-orders.

21.1 GENERAL APPROACH

When analyzing community data with an ordination method such as principal components analysis, a typical pattern observed is an "arch" to the SUs (e.g., Figure 19.6). Such distortions can cause great difficulties with ecological interpretations.

Many of the ordination techniques in common use, such as polar ordination (PO, Chapter 18), principal components analysis (PCA, Chapter 19), and

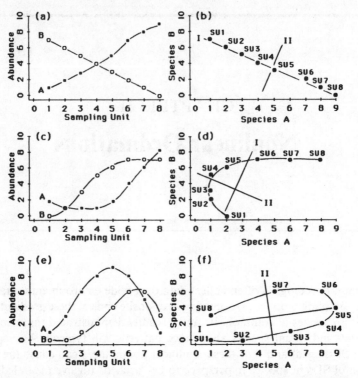

Figure 21.1 *Abundances of two species, A and B, in eight sampling units (SUs) located at fixed intervals in a transect placed along a single environmental gradient. (a) Typical abundance patterns for narrow gradients or wide amplitude species and (b) SUs in species space and the fit of PCA axes I and II in this space. (c) Typical abundance patterns for medium gradients or amplitudes and (d) SUs in species space and the fit of PCA axes I and II in this space. (e) Typical abundance patterns for wide gradients or narrow amplitude species and (f) SUs in species space and the fit of PCA axes I and II in this space.*

correspondence analysis (COA, Chapter 20) assume linear or near-linear relationships between species responses in SUs sampled along an environmental gradient. This assumption is reasonable when the underlying environmental gradient is narrow or the species have broad ecological amplitudes (Noy-Meir and Whittaker 1977). Take, for example, the near-linear relationship between species *A* and *B* sampled in eight SUs along a hypothetical gradient shown in Figure 21.1*a*. In this case, the first axis (I) of a PCA (or a COA) will fit the corresponding SU pattern very well and, thus, axis

I will be an accurate reflection of the underlying environmental gradient (Figure 21.1*b*).

If the environmental gradient is wider, yet not extreme, the typical species abundance patterns for species *A* and *B* may show some overlapping of segments of their individual bell-shaped distribution curves along the gradient (Figure 21.1*c*). In this case, the abundance relationship between the two species is clearly nonlinear, and the corresponding SU pattern will be an arching curve, as shown in Figure 21.1*d*. The first two axes of a PCA will reflect this arching, but within the centered and rotated PCA coordinate system and, thus, require *two* axes to represent the *one* underlying gradient. If this arching trend could be removed (or detrended) by forming a single axis to reflect the single environmental gradient, this could lead to easier ecological interpretations.

Phillips (1978) described a method using a second-order polynomial regression for unfolding or detrending the arched curve resulting from the first two principal components of a PCA on moderately nonlinear data sets, a procedure he called *polynomial ordination* or what we call *detrended principal components* (DPC). Phillips' method is a relatively simple procedure for overcoming the arch problem of PCA. In Section 21.2.1, we describe DPC.

As evident from our descriptions in the previous chapter (e.g., Figure 20.4), COA also has the arch problem (also see Gauch 1982). Hill and Gauch (1980) have described *detrended correspondence analysis* (DCA) for detrending the SU arch often observed with COA. DCA uses a complex algorithm for removing the arch (Austin 1985); we therefore have chosen to present the simpler DPC procedure in this chapter. However, we recommend that the student using COA should also consider using DCA if an arch is evident.

If the underlying environmental gradient is very wide (e.g., extending from the bottom to the top of a mountain), species abundances over such a broad gradient are likely to be more completely overlapping (e.g., Figure 21.1*e*). The overlapping of such bell-shaped species curves result in SU patterns that appear as "loops" (Pielou 1984), "closed loops" (Pielou 1977), or "spirals" (Noy-Meir and Austin 1970). A PCA on such strongly nonlinear data will be unsatisfactory (see Figure 21.1*f*) since PCA cannot open such loops or spirals to reflect the single underlying gradient. In fact, DPC will probably not be satisfactory either. With such acute nonlinear data structures, various nonlinear ordination methods, such as *continuity analysis* (Noy-Meir 1974), *Gaussian ordination* (Gauch et al. 1974), or *nonmetric multidimensional scaling* (NMDS, Anderson 1971) must be used. In Section 21.2.2 we describe NMDS as a method that more effectively recovers strong nonlinear data structures than do other commonly used ordination methods (Kenkel and Orloci 1986; Minchin 1987).

21.2 PROCEDURES

21.2.1 Detrended Principal Components (DPC)

The following method of DPC is basically an adaptation and renaming of nonlinear factor analysis described by McDonald (1962, 1967), who used it on psychometric data (Phillips 1978). We illustrate DPC as a simple nonlinear ordination procedure because we suspect *moderately* nonlinear data structures are common in many community ecology studies; certainly there are numerous examples that seem to bear this out (see Gauch 1982). Also, DPC examines nonlinearity as a logical follow-up to PCA, that is, if an arching SU pattern exists as a result of moderately nonlinear species abundance relationships, then the DPC procedure can test for the presence and significance of arching SUs within a PCA and, if significant, detrend the arch into one axis to reflect the underlying environmental gradient.

STEP 1. PERFORM PCA ON THE DATA. The first procedure for DPC is to perform a PCA on the data (Chapter 19). The results of this PCA will be an SU ordination based on the SU coordinates from the eigenvalues and eigenvectors of the R matrix (see Section 19.2).

STEP 2. REGRESS PCA COMPONENT II ONTO COMPONENT I. A parabolic (or second-order) polynomial regression of principal component II onto principal component I is determined. The polynomial regression equation is

$$Y = B_0 + B_1 X + B_2 X^2 \tag{21.1}$$

where Y is the SU coordinates on PC II (the dependent variable), X is the SU coordinates on PC I (the independent variable), and the B's are parameters that define the shape of the parabolic curve. B_0 is the intercept or position parameter and B_1 and B_2 define the openness and direction of the parabola. (If B_2 is positive, the parabola opens upward; if B_2 is negative, the parabola opens downward.) Multiple linear regression is used to estimate the B's in Eq. (21.1); we refer the student to Sokal and Rohlf (1981, Box 16.1) for these computations.

STEP 3. TEST FOR SIGNIFICANCE OF THE REGRESSION. The parabolic regression is now tested for significance by using the coefficient of multiple determination, R^2 (see Sokal and Rohlf 1981, Box 16.2) to estimate an F ratio:

$$F = \frac{R^2/2}{(1 - R^2)/(N - 2 - 1)} \tag{21.2}$$

where N is the number of SUs. If the parabolic regression is significant

$(P = 0.05;$ df $= 2,$ $N - 2 - 1)$, then the original PCA axes I and II can be combined into a single detrended principal component axis. This single DPC axis should better reflect the underlying environmental gradient responsible for the arching in PCA.

STEP 4. CONSTRUCT THE DPC AXIS. A single DPC axis is constructed by projecting the position of the SUs in the two-dimensional PCA system perpendicularly onto the parabolic curve. This is most easily done graphically by eyeing the results. We arbitrarily set the DPC I axis to range from -1 to $+1$ with the SUs on the ends representing the extremes (see example in Section 21.3.1). This works well with simple data sets, but for more complex data sets a more precise algebraic method is provided by Phillips (1978). The parabola is thus detrended by forming a single DPC ordination axis.

21.2.2   Nonmetric Multidimensional Scaling (NMDS)

The following procedures for an NMDS ordination are based on those described by Fasham (1977), Pimentel (1979), and Prentice (1977), as developed in psychometrics (see Shepard 1980). In those cases of strongly nonlinear relationships among species and the resultant SU loop or spiral in multivariate space (e.g., Figure 21.1f), NMDS is an informative ordination method; some forms of NMDS appear to be superior to others (Kenkel and Orloci 1986).

STEP 1. RANK-ORDER SU PAIRS. First, compute a Q-mode resemblance measure between all pairs of SUs (such as any of the distance functions in Chapter 14). Then these measures are ranked in ascending order, that is, from lowest to highest dissimilarity (ds) as

$$\text{ds}_{i,j} < \text{ds}_{k,l} \tag{21.3}$$

between SU pairs $i < j$ and $k < l$. Kenkel and Orloci (1986) compared ordination methods using simulated data and found the most robust NMDS strategy was to first standardize the data by SU; chord distance [Eq. (14.9)], for example, implicitly does this.

Because NMDS only uses the nonmetric rank-orders in subsequent steps (rather than the actual Q-mode measures), it is often stated that the selection of a resemblance measure for ranking is relatively unimportant with NMDS. While NMDS may be less sensitive to the measure selected than other ordination methods, the point is that the chosen measure should allow a scaling method (like NMDS) to correctly order SUs along a *straight* line reflecting an underlying habitat gradient. This is a rather complex issue, which we discuss further in Section 21.6.

STEP 2. RANK-ORDER SU ORDINATION DISTANCES. Using the coordinates for SUs from a previous two- or three-dimension ordination, we compute the distances between all pairs of SUs. Then we rank-order these distances (d) in ascending order, that is,

$$d_{i,j} < d_{k,l} \qquad (21.4)$$

between SU pairs $i < j$ and $k < l$. Any of the distance measures in Chapter 14 could be used and, in fact, the measure need not be Euclidean (Fasham 1977).

The objective here is to come up with an initial configuration that is based on a reasonably good ordination result (such as a PO, PCA, or COA), so that the optimization algorithm used in Step 3 finds a "global" minimum, not some "local" minimum (Fasham 1977). To explain what this jargon means, we return to an analogous problem in Chapter 7. To estimate values for the parameters of the lognormal model [Eq. (7.3)], we described several methods for coming up with reasonable initial estimates; obviously, the better the initial estimates were, the fewer the number of iterations required to minimize the differences between observed and expected frequencies. The problem here is similar; the better the initial estimates (the ordination rank-orders), the fewer iterations required to minimize (in Step 3 below) the differences in the ranks between the SU × SU distances (Step 1) and the ordination distances (Step 2). Poor initial estimates can sometimes result in "local" minimums and ineffective ordinations; good initial estimates will result in "global" minimums.

If the rank-order of SU resemblances from the community data perfectly match the SU distances from the ordination (as a monotonically increasing function, Step 3), then the initial configuration would be considered a satisfactory NMDS ordination. Basically, NMDS assumes that a good ordination has an optimal rank-order relationship between SU by SU distances for the data and the ordination space. However, a perfect match is unlikely (except for very small data sets, Section 21.3), and we proceed to Step 3.

STEP 3. OPTIMIZE RANK-ORDERS. An iterative algorithm is set up by first computing a monotonically increasing curve that gives a "best fit" to the relation between the data SU by SU resemblances, $ds_{i,j}$, and the ordination SU by SU distances, $d_{i,j}$, using Kruskal's nonparametric regression technique. For each SU pair, i and j, the regression estimates a value $\hat{d}_{i,j}$ for each value of $ds_{i,j}$. These estimated values are *not* distances, but a set of numbers known to be monotonic with the values of $ds_{i,j}$. They are used to compute a measure of departure from monotonicity. Kruskal (1964) defined such a measure, called stress (SR), as

$$SR = \sqrt{\frac{\sum_{i<j}^{M}(d_{ij} - \hat{d}_{ij})^2}{\sum_{i<j}^{M}(d_{ij}^2)}} \qquad (21.5)$$

where the summation is over all SU(i, j) pairs, $i < j$, where $M = [N(N - 1)/2]$. The smaller the value of SR, the closer the relation is approaching monotonicity, that is, low values of SR indicate a high degree of concordance of the rank-orderings of the ordination SU by SU distances with the data SU by SU resemblances.

Iteratively, the NMDS algorithm moves the SUs within ordination space in order to minimize stress. This minimization process, called the *method of steepest descent*, is described by Kruskal (1964). The mathematical details are beyond the scope of this book, but the process is built into our program NMDS.BAS (see accompanying disk).

STEP 4. CONSTRUCT THE NMDS ORDINATION.   After iterations fail to significantly move the SUs in the configuration, the final ordination is formed by normalizing this configuration (i.e., to zero centroid and unit dispersion). However, if these NMDS axes (as defined by the SU coordinates) are strongly correlated, they can be rotated to uncorrelated principal axes (e.g., as in PCA; see Chapter 19) and the resultant rotated axis coordinates used to construct the final NMDS ordination. This rotation option is provided in program NMDS.BAS.

21.3   EXAMPLE: CALCULATIONS

21.3.1   Detrended Principal Components (DPC)

To illustrate a DPC ordination on ecological data, the results of PCA on the data for two (S) species in five (N) SUs from Table 19.1 will be used.

STEP 1. PERFORM PCA ON THE DATA.   The scores for the five SUs on principal components (PCs) I and II are given in Table 21.1.

TABLE 21.1 *Regression variables as coordinates for five SUs on two principal components (PC I and PC II) along with estimates for means, standard deviations (SD), correlations (r), regression coefficients (B's), and the coefficient of determination (R^2)*

	SUs						
Variable	(1)	(2)	(3)	(4)	(5)	Mean	SD
Y = PC II	−0.28	−0.11	+0.12	−0.50[a]	+0.28	−0.10	0.31
X = PC I	−0.62	+0.55	+0.78	0.00	−0.72	0.00	0.67
X^2 = (PC I)^2	+0.38	+0.30	+0.61	0.00	+0.52	0.36	0.23

$$r(Y, X) = 0.00, \quad r(Y, X^2) = 0.88, \quad \text{and} \quad r(X, X^2) = 0.05$$

$$B_0 = -0.52, \quad B_1 = -0.018, \quad B_2 = 1.17, \quad \text{and} \quad R^2 = 0.77$$

[a] In order to more effectively illustrate the fitting of a parabolic regression to principal components I and II, we have arbitrarily positioned SU 4 at 0.0 and −0.5, instead of at 0.0 and 0.0 as shown in Chapter 19, Figure 19.4.

STEP 2. REGRESS PCA COMPONENT II (Y) ONTO COMPONENT I (X).   First compute estimates for means, standard deviations, correlations, and regression coefficients as outlined in Sokal and Rohlf (1981, Box 16.1 and 16.2). These estimates are provided in Table 21.1. The resultant regression is [Eq. (21.1)]

$$Y = -0.52 + (-0.018X) + (1.17X^2)$$

STEP 3. TEST FOR SIGNIFICANCE OF THE REGRESSION.   Using the coefficient of multiple determination $R^2 = 0.77$, compute the F ratio [Eq. (21.2)]:

$$F = (0.77/2)/[(1 - 0.77)/(5 - 2 - 1)] = 3.55$$

This F statistic is not significant ($P = 0.05; df = 2, 2$). With this worked example, the statistical insignificance is due to the low degrees of freedom; the parabola fits the data very well (see Figure 21.2a).

Normally, if insignificance is shown at this step, it is assumed that either (1) the linear PCA is an adequate reduction of the data or (2) a strong nonlinear data structure exists and the loop or spiral cannot be unfolded. However,

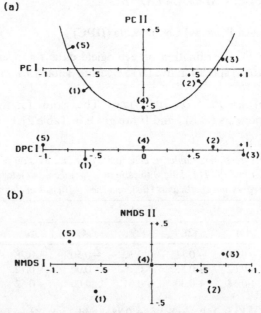

Figure 21.2   *Ordination of five SUs: (a) principal components analysis showing positions on components I (PC I) and II (PC II) with SU 4 arbitrarily positioned at (0.0, −0.5), and the fit of a second-order polynomial regression of PC II onto PC I is shown along with the detrended principal component axis I (DPC I); (b) nonmetric multidimensional scaling showing positions on two NMDS axes.*

for illustrating the calculations, we assume the regression is significant and proceed to Step 4.

STEP 4. CONSTRUCT THE DPC AXIS.   The single DPC axis is constructed by a perpendicular projection of the SUs onto the parabolic curve (Figure 21.2a). The extreme SUs (5 and 3) are designated values of -1 and $+1$, respectively, on the DPC axis and the remaining SUs are projected onto this axis by "eyeing" their relative positions to each other. Note how the arched pattern of the five SUs on PC I and PC II is now represented by a single, unfolded DPC axis.

21.3.2   Nonmetric Multidimensional Scaling (NMDS)

Again, as an example, we use the data and results from the PCA on the contrived data matrix of two species in five SUs in Table 19.1.

STEP 1. RANK-ORDER SU PAIRS.   A Q-mode resemblance measure, Euclidean distance [ED, Eq. (14.1)] was computed for all (i, j) SU pairs, and these were ranked from lowest to highest dissimilarity [Eq. (21.3)]:

Rank: 1   2   3   4   5   6   7
         8   9   10
Pair: $ds_{2,3} < ds_{2,4} \leq ds_{1,5} \leq ds_{1,4} < ds_{3,4} < ds_{4,5} < ds_{1,2}$
         $< ds_{1,3} < ds_{2,5} < ds_{3,5}$
$ED_{i,j}$:   1.00 < 2.24 ≤ 2.24 ≤ 2.24 < 2.83 < 3.16 < 4.24
         < 5.00 < 5.39 < 5.83

STEP 2. RANK-ORDER SU ORDINATION DISTANCES.   From the PCA on these data, the coordinates for the first two principal components (Table 19.3b) are used to compute ED distances between all (i, j) SU pairs for this initial configuration. These SU by SU distances are then ranked from lowest to highest distance [Eq. (21.4)]:

Rank: 1   2   3   4   5   6   7
             8   9   10
Pair: $d_{2,3} < d_{2,4} < d_{1,5} < d_{1,4} < d_{4,5} < d_{3,4} < d_{1,2}$
             $< d_{2,5} < d_{1,3} < d_{3,5}$
$ED_{i,j}$:   0.33 < 0.56 < 0.57 < 0.68 < 0.77 < 0.79 < 1.18
             < 1.33 < 1.46 < 1.51

Note that these ranks for ordination distances between SU pairs do *not* quite match the rank-order for the SU by SU dissimilarities from the data (Step 1). The orders for SU pairs (4, 5) and (3, 4) and for (2, 5) and (1, 3) are reversed.

STEP 3. OPTIMIZE RANK ORDERS.   To achieve the correct ranking, the distance between SU 1 and SU 3 in the ordination configuration should be less

TABLE 21.2   *A final configuration (coordinates) for an NMDS ordination after iterations to move SUs within an initial PC I and II configuration to achieve concordance between the rank orders of SU by SU distances for the data [ds, Eq. (21.3)] and the ordination [d, Eq. (21.4)]*

	Initial[a]		Final	
SU	I	II	I	II
(1)	−0.62	−0.28	−0.57	−0.28
(2)	+0.55	−0.11	+0.60	−0.11
(3)	+0.78	+0.12 ⟹	+0.73	+0.12
(4)	0.00	0.00	0.00	0.00
(5)	−0.72	+0.28	−0.78	+0.28

[a] see Table 19.3.

than the distance between SU 2 and SU 5. We can simply move SUs 1 and 3 closer together and SUs 2 and 5 farther apart by changing their coordinates in the initial configuration. If we add or subtract 0.05 units to the positions of SUs 1, 2, 3, and 5 on the first ordination axis (Table 21.2) and then recompute SU by SU distances for this new ordination configuration, the resultant rank-orders are

Rank: 1   2   3   4   5   6   7
          8   9   10

Pair: $d_{2,3} < d_{2,4} < d_{1,5} < d_{1,4} < d_{3,4} < d_{4,5} < d_{1,2}$
        $< d_{1,3} < d_{2,5} < d_{3,5}$

$ED_{i,j}$: $0.26 < 0.60 < 0.61 < 0.64 < 0.74 < 0.83 < 1.18$
              $< 1.36 < 1.43 < 1.52$

Note that this $d_{i,j}$ rank-order is now in complete concordance with the $ds_{i,j}$ rank-order in Step 1.

For this example, we simply used our intuition to adjust coordinates. The NMDS ordination algorithm would have computed stress [SR, Eq. (21.5)] by using nonparametric regression and then moved the SUs within the configuration (adjusted coordinates) by taking the partial derivatives of SR with respect to each dimension to obtain a "negative gradient" for each point and then moved the points in the configuration "down" this gradient with what is known as a stepping function (steepest descent). As noted earlier, these mathematical procedures are beyond the scope of this book.

STEP 4. CONSTRUCT THE NMDS ORDINATION.   The above configuration (Step 3) is accepted as the final NMDS ordination (Figure 21.2*b*). Note the expected close similarity between this ordination and that for the original PCA ordination (Figure 19.4). We discuss the effect of initial configuration on the final NMDS ordination in Section 21.6.

TABLE 21.3 *Results for detrended principal components using program DPC.BAS on the pattern of six Panamanian localities on principal components I and II*

		Statistics		
Variable	Mean	Variance	Standard Deviation	Standard Error
X = PC I	-0.000	0.605	0.778	0.318
X^2 = (PC I)^2	0.504	0.757	0.870	0.355
Y = PC II	-0.000	0.235	0.485	0.198

		Correlations		
Comparison Variables	Correlation Coefficient	Standard Error	Degrees Freedom	t Statistic
PC I, (PC I)^2	-0.922	0.193	4	-4.769
PC I, PC II	. 0.000	0.500	4	0.001
(PC I)^2, PC II	0.252	0.484	4	0.521

Second-order polynomial regression equation

$$Y = -0.474 + (0.971X) + (0.941X^2)$$

Coefficient of multiple determination $R^2 = 0.426$
F ratio = 1.115 at degrees of freedom = (2, 3)

21.4 EXAMPLE: PANAMANIAN COCKROACHES

Using the results obtained for a PCA on the Panamanian cockroach data (Table 19.4), the pattern of the six SUs on PC I and PC II (Figure 19.5) was tested for arching and, thus, for possible detrending by running a DPC analysis. The BASIC computer program DPC.BAS (accompanying disk) was used. The results (Table 21.3) suggest that although there is a slight parabolic curvature to the pattern of the six localities or SUs on PC I and II (Figure 21.3a), it is not significant. The clustering of localities 3, 4, and 5 strongly influences the shape of the parabola.

For a NMDS ordination, the cockroach data were again used. Chord distances [Eq. (14.9)] were computed as the measure of resemblances between SU pairs. An initial ordination configuration from a PCA on these data (Table 19.4) was used in program NMDS.BAS to produce a final rotated configuration (Figure 21.3b) after a series of iterations reduced stress to a very low value (i.e., the rank-orders of the final ordination distances very closely matched the rank-order of the data SU by SU resemblances).

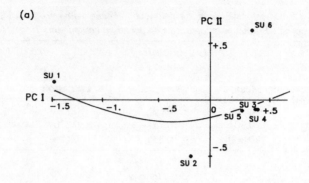

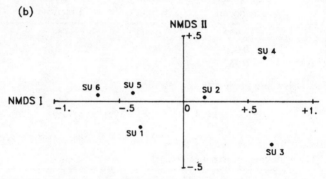

Figure 21.3 *Ordination of six Panamanian cockroach localities (SUs): (a) principal components analysis showing fit of the second-order polynomial regression of PC II onto PC I, and (b) nonmetric multidimensional scaling of SUs on two NMDS axes, I and II.*

Note that this NMDS ordination is substantially different from that for PCA (Figure 19.5), but somewhat similar to that for COA (Figure 20.3). This difference is probably related to the fact that we used chord distance to obtain the rank-order of SU resemblances (ds), that is, the rank-ordering that the NMDS algorithm is attempting to match by moving SUs within a configuration. Recall that chord distance implicitly standardizes the data by SU (column), whereas PCA is based on a scalar product or correlation by species, that is, a row standardization. On the other hand, COA uses a double (column and row) standardization of the data and, thus, our COA ordination configuration is more similar to that of our NMDS. Kenkel and Orloci (1986) demonstrated that a NMDS on SU (column) standardized data (e.g., chord distance) was the most robust strategy for recovering various degrees of nonlinearity caused by species "turnover" along simulated gradients.

TABLE 21.4 *Results for detrended principal components using program DPC.BAS on the pattern of ten forest sites on principal components I and II*

		Statistics		
Variable	Mean	Variance	Standard Deviation	Standard Error
$X = $ PC I	0.000	0.497	0.705	0.223
$X^2 = $ (PC I)^2	0.447	0.054	0.232	0.073
$Y = $ PC II	0.000	0.153	0.391	0.124

		Correlations		
Comparison Variables	Correlation Coefficient	Standard Error	Degrees of Freedom	t Statistic
PC I, (PC I)^2	−0.111	0.351	8	−0.315
PC I, PC II	0.000	0.354	8	0.001
(PC I)^2, PC II	0.833	0.196	8	4.259

Second-order polynomial regression equation

$$Y = -0.635 + (0.051X) + (1.419X^2)$$

Coefficient of multiple determination $R^2 = 0.702$
F ratio $= 8.264$ at degrees of freedom $= (2, 7)$

21.5 EXAMPLE: WISCONSIN FORESTS

The PCA ordination of 10 upland forest sites, southern Wisconsin (Table 19.5, Figure 19.6) was examined for significant arching of the SUs by using the BASIC program DPC.BAS. The results from this program are given in Table 21.4. The polynomial regression is significant ($P = 0.05$; df $= 2, 7$) and the DPC axis obtained by projecting, perpendicularly, the 10 SUs onto the parabola is shown in Figure 21.4a. The results for this DPC analysis are very similar to the results for a detrended correspondence analysis on the same 10 upland forest sites (Gauch 1982, p. 158), although our data matrix was simplified to only 8 species from the 14 used by Gauch (1982, p. 122).

A NMDS ordination of this same forest data (again using chord distance and a PCA initial configuration with program NMDS.BAS) resulted in an unrotated pattern of SUs (Figure 21.4b) that is clearly less arched than either the SU pattern for a PCA (Figure 19.6) or for a COA (Figure 20.4). In other words, the NMDS algorithm repositioned the 10 SUs within the ordination configuration to get their rank-order distances to as closely match as possible

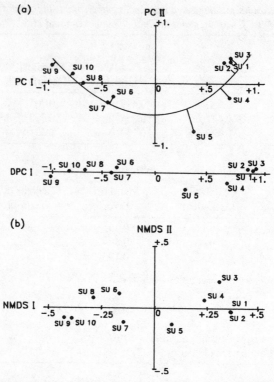

Figure 21.4 *Ordination of 10 upland forest sites (SUs): (a) polynomial ordination (DPC) on PC I and PC II and the detrended principal component axis (DPC I) formed by the projection of these SUs onto the parabola, and (b) nonmetric multidimensional scaling showing positions of the SUs on two NMDS axes, I and II.*

the rank-order of the SU by SU chord distances. The repositioning stabilized the SUs into a configuration where the first ordination axis (NMDS I) very closely approximates a straight line through a forest gradient from SUs 1 and 2 to SUs 9 and 10 (see Gauch 1982 and Peet and Loucks 1977).

21.6 ADDITIONAL TOPICS IN NONLINEAR ORDINATION

As discussed in Chapter 17, the aim of ordination is to position SUs in species hyperspace such that the spatial axes represent underlying environmental gradients, which hopefully lead to meaningful ecological interpretations (Chapter 24). Problems arise when species relationships along these gradients are nonlinear (e.g., overlapping bell-shaped curves). The derived ordination axes do not give a true representation of the underlying gradients (e.g., Figure 21.1) and gradients can end up as in-curved arches (two-dimensions) or spirals (three-dimensions). In this chapter we presented a procedure for detrending principal components (DPC) that "unfolds" an arch. We then presented nonmetric multidimensional scaling (NMDS) as a procedure for "rank-ordering" an arch or spiral. However, it is clear that the arch/spiral problem really goes back to the failure of resemblance measures to estimate correctly ecological distance, that is, the real ecological gradient distance between SUs. This failure is particularly acute when SUs have few or no species in common (i.e., occur at opposite ends of a wide gradient) and/or when species relations are nonlinear (again, typically along a wide gradient). The early development of ordination methods to account for this failure were somewhat successful, but also limited (see Gauch 1973, Gauch et al. 1974, Ihm and Groenewoud 1975). However, the search continues for ecologically appropriate resemblance measures for use with ordination (Bradfield and Kenkel 1987, Faith et al. 1987, Fewster and Orloci 1983), and we advise the student to be on the lookout for new developments in nonlinear ordination methodology, including those based on fuzzy set theory (Roberts 1986).

In this chapter we chose to present the DPC procedure for detrending the arch often found in the display of SUs within the coordinates formed by principal components I and II. We also note that the presence of an arch between other combinations of components (e.g., PC I versus PC III, or PC II versus PC III) can be examined using this procedure, as shown by Phillips (1978).

We also presented NMDS as an ordination method that is proving to be a robust strategy for dealing with nonlinearities resulting from bell-shaped (Gaussian) curves of species along steep or long gradients (Kenkel and Orloci 1986). Such curves are evident for species along a steep topographic moisture gradient in the Santa Catalina Mountains, southern Arizona (Niering and Lowe 1984, Whittaker and Niering 1964). Other ordination methods developed to deal with such overlapping bell-shaped curves include the *Gaussian ordination method* (Gauch et al. 1974), where SUs are arranged along the first ordination axis such that the pattern of each species along this axis best fits a Gaussian curve. The procedure begins with an initial positioning of the SUs to approximate bell-shaped curves for the species and then the SU positions are iteratively adjusted to obtain the best fit of all species to the Gaussian model.

Ihm and Groenewoud (1975) and Johnson and Goodall (1979) describe similar procedures based on the Gaussian model. As a modification of this, Orloci (1980) suggested an ordination algorithm that allows the inclusion of other species-abundance models other than Gaussian. The basic Gaussian ordination method has been effective when dealing with data having a single dominant gradient, but attempts to extend the method to multiaxes have generally been unsatisfactory (Gauch 1982).

A nonlinear ordination method related to NMDS has been described by Noy-Meir (1974). He called the procedure *continuity analysis* because SUs are iteratively ordered in continuous sequences in such a way as to optimize what Noy-Meir termed their *local similarities*. The procedure was derived from a psychometrics procedure known as *parametric mapping* (Shepard and Carroll 1966). However, continuity analysis appears to only work well with data sets of moderate nonlinearities (Gauch 1982), where the species-abundance curves are only partial segments of bell-shaped curves (e.g., Figure 21.1c).

Bradfield and Kenkel (1987) introduced another scheme for dealing with nonlinear ordination. Their method, called *flexible shortest path adjustment*, involves replacing certain values in the SU by SU distance matrix. These replacements are termed "shortest paths," and represent recalculated distances between SUs with little or no overlap in species composition. The advantage of this technique is that the user has the flexibility to determine conditions whereby shortest paths are calculated in order to achieve the most efficient ordination.

The NMDS algorithm begins with an initial configuration of SUs within a coordinate system and then SUs are moved within this configuration so that the rank-order of SU by SU distances matches as closely as possible the rank-order of SU by SU resemblances from the data matrix. It has been suggested by some comparative studies that the choice of initial configuration is not critical as the algorithm normally converges to an optimal solution (Fasham, 1977). However, our experience suggests that local optima are highly likely with small data sets and even quite possible with large data sets. Thus, we recommend that a number of different initial configurations be used, including random configurations (Kenkel and Orloci 1986). Fasham (1977) recommended at least two different initial configurations be used: a random pattern and another based on COA ordination. The aim is to obtain similar, not identical, final SU configurations using different initial configurations with NMDS. (Recall that the method is only based on optimizing rank-orders, not actual metric distances.) Different computer programs for NMDS produce final configurations that differ in scale and rotation.

The NMDS algorithm also optimizes with respect to the number of axes or dimensions used in the configuration, that is, the SU pattern for a two-dimensional NMDS ordination will be different from the SU pattern on the first two dimensions of a three-dimensional NMDS ordination. Thus, the number of important ordination axes (i.e., the number of significant underlying gradients) must be known or specified *a priori*. Of course, the ecologist who collected the community data is most likely to be able to specify this. If uncertain, an NMDS can be repeated using higher or lower dimensions, with the final solution chosen based on the one that provides the best ecological interpretation.

Of course, evaluating and comparing all of the different ordination methods is beyond the scope of this book. The student is referred to the excellent summaries by Gauch (1982), Minchin (1987), Orloci (1978), and Pielou (1984). Pielou (1984) states that different ordination methods "have merit in appropriate circumstances." This statement emphasizes that the selection of an ordination method (or methods) depends greatly on the character of the data set in hand and the objectives in mind.

21.7 SUMMARY AND RECOMMENDATIONS

1. Community data obtained by sampling along a long, steep gradient typically include bell-shaped species-abundance curves with nonlinear overlaps between species, resulting in SU patterns within ordinations that are strongly arched, looped, or spiraled (Section 21.1).

2. When a principal components analysis (PCA, Chapter 19) demonstrates an arched patterning in the SUs within the ordination, this arch can be detrended or unfolded using polynomial regression, a procedure called *polynomial ordination* (Phillips 1978) or *detrended principal components* (DPC) analysis. We recommend this procedure as a logical follow-up to PCA (Section 21.2.1).

3. When a correspondence analysis (COA, Chapter 20) shows the SU arching pattern resulting from nonlinear relationships between species, we recommend the use of detrended correspondence analysis (DCA) as a method for straightening the arch (Hill and Gauch 1980). Although the algorithm for DCA is quite complex and beyond the scope of this text, computer programs are available (e.g., the Cornell Ecology Programs; see Gauch 1982, p. 256).

4. When SU patterns within an ordination are likely to be strongly nonlinear (looped or spiraled), we recommend ordination by nonmetric multidimensional scaling (NMDS) (Section 21.2.2). Comparative studies support a NMDS procedure that uses a Q-mode resemblance measure [ds, Eq. (21.3)] that normalizes the data by SU (e.g., chord distance, Orloci and Kenkel 1986, or percent dissimilarity, Minchin 1987).

5. Current ecological resemblance measures (e.g., distances or similarities; Chapter 14) fail to adequately measure the true separation of SUs located at opposite ends of a gradient (Section 21.6). The search goes on for measures to improve the performance of ordination (e.g., Bradfield and Kenkel 1987).

6. The number of dimensions of a given NMDS ordination must be specified before a given analysis, as the NMDS algorithm depends on a fixed dimensionality (Section 21.6). If there is uncertainty as to the number of important underlying gradients (the usual case), we recommend that the NMDS be repeated and the final dimensionality selected on the basis of ease of ecological interpretability.

PART SEVEN
COMMUNITY INTERPRETATION

CHAPTER 22

Background

In the nine preceding chapters, we presented various methods of community classification and ordination. Basically, the objective of classification is to reduce a data matrix of S species and N sampling units (SUs) into $g < N$ groups, where the SUs within groups are more similar to each other than between groups, while the objective of ordination is to transform the S by N data matrix into a reduced $k < S$ coordinate system (Green 1980). In other words, the aim of classification is to cluster or group SUs into homogeneous groups, while the aim of ordination is to order SUs within an axis system to reflect similarity patterns. The actual underlying environmental factors that might be responsible for the resulting classification or ordination patterns did not directly enter into our analyses, although a consideration of these factors would enter into the choice of a specific method. In the following two chapters, we present procedures for exploring the influence of environmental factors on observed patterns in community classification (Chapter 23) and ordination (Chapter 24).

Given the g groups or communities resulting from a classification, the environmental interpretation of these groups logically entails the use of some statistical method for testing differences between environmental factors (variables) measured within each of the g groups. These methods may include, for example, nonparametric rank and sign tests or parametric one-way analysis of variance and mean comparison tests. Also available to the ecologist, who constantly deals with many simultaneous factors, are multivariate statistical methods. In particular, discriminant analysis is a powerful family of procedures for examining many factors simultaneously in order to examine community differences (Chapter 23). We distinguish two types of discriminant analysis:

simple and multiple. The former is used to compare pairs of groups and the latter compares all g groups simultaneously. In Chapter 23 we present the procedures for a simple discriminant analysis (SDA).

Given N SUs ordered along $k < S$ axes (gradients) resulting from an ordination, the environmental interpretation of this ordination logically entails the use of methods for relating these axes to environmental variables measured in each of the SUs. These methods include the use of simple and multiple linear regression to relate ordination axes to environmental factors (Chapter 24).

22.1 MATRIX VIEW

Community interpretation may be viewed as a two-step process. Initially, data on species presence–absence or abundances in SUs are used to classify or ordinate communities. Next, using data on environmental factors (e.g., soil water and pH) measured within the SUs, an interpretation of community patterns is attempted (Figure 22.1). Note that we are now using the entire data matrix, that is, all rows and columns of the species subset and all rows and columns of the environmental factor subset.

22.2 SELECTED LITERATURE

Simple and multiple discriminant analysis have been used to provide an environmental interpretation for a wide range of communities—streams, deserts, and forests (Table 22.1). One of the very first ecological uses of discriminant analysis was in a beech forest, using soil data for an interpreta-

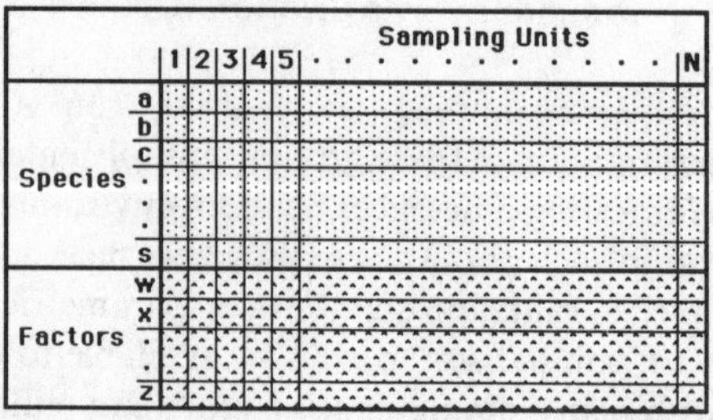

Figure 22.1 *The shaded area indicates the form of the ecological data matrix for studies of community interpretation. Of interest is the species-SU data for use in classification and ordination and the use of environmental factor data contained in the SUs.*

TABLE 22.1 *Selected literature giving examples of community interpretation studies using simple discriminant analysis (SDA), multiple discriminant analysis (MDA), and multiple linear regression (MLR)*

Location	Community	Method	Reference
Montana	Blue grouse	SDA	Martinka 1972
Virginia	Woodpeckers	SDA	Conner and Adkisson 1976
Alaska	Salmon streams	SDA	Swanston et al. 1977
England	Beech forest	MDA	Norris and Barkham 1970
New York	Benthos	MDA	Walker et al. 1979
West Germany	Macrophytes	MDA	Wiegleb 1981
Arizona	Desert birds	MDA	Rice et al. 1983
Po Delta, Italy	Oak forest	MDA	Gerdol et al. 1985
California	Redwood forest	MLR	Waring and Major 1964
Wyoming	Alpine tundra	MLR	Scott and Billings 1964
New Brunswick, Canada	Forest	MLR	Forsythe and Loucks 1972
Ngari, Tibet	Plateau–montane	MLR	Chang and Gauch 1986

tion (Norris and Barkham 1970). The use of multiple linear regression to interpret ordination results has also been used by ecologists, but largely in terrestrial communities.

CHAPTER 23

Classification Interpretation

In this chapter we present simple discriminate analysis (SDA), a method that tests for significant differences in abiotic characteristics among communities delimited by classification. SDA enables us to test for significant differences among communities by computing a multivariate distance (D^2) statistic and an F ratio. In addition, we can compute a discriminant function that is useful for evaluating the relative contribution of each variable to the environmental discrimination among communities.

23.1 GENERAL APPROACH

Gauch (1982) lists three purposes for performing a classification in ecology: (1) to summarize large, complex sets of data; (2) to aid in the environmental interpretation of patterns of community variation; and (3) to refine models of community structure. The emphasis of this chapter is on the environmental interpretation of community variations.

In Part IV (Chapters 15 and 16), association and cluster analyses were presented for classifying a collection of SUs into groups or clusters. Recall that the classification process can be viewed as a reduction of a data matrix of S rows (species) and N columns (SUs) into $g < N$ groups, where the SUs within each of the g groups are more similar to each other than SUs between groups (Green 1980). A common goal in community ecology is to identify g *homogeneous* communities from samples taken over diverse environments. Given a data set with this $g < N$ structure, we can now get down to questions dealing with ecological interpretation. For example, are there environmental

TABLE 23.1   *Parametric and nonparametric statistical tests useful in comparing groups (communities) for differences in both qualitative and quantitative environmental factors*

Number of Groups	Test Type	Test Name	Comment
Two	Parametric	t-statistic	Widely used univariate test of mean differences
	Parametric	Hotelling T^2	Multivariate equivalent to the t-statistic
	Parametric	Simple discriminant analysis	F-test is related to Hotelling's T
	Nonparametric	Sign test	Applicable to qualitative, paired data
	Nonparametric	Wilcoxon	Quantitative unpaired data; tests differences in ranks
Three or More	Parametric	Analysis of variance, one-way with mean contrasts	Widely used univariate test of overall differences and group mean comparisons
	Parametric	Multivariate discriminant analysis	Multivariate test of mean vector differences and multivariate distance

factors present that, because of differential species responses, give rise to the g groups? If so, what does this imply ecologically?

To test for significant differences among communities with respect to a set of environmental factors measured in each of the SUs in g groups, the appropriate statistical test must be selected. In Table 23.1, a listing of some of the many possible tests is given. Although not intended to be exhaustive, this table lists some important tests for ecological applications.

Parametric tests have a number of underlying assumptions related to the statistical estimates of the population parameters being compared. For example, it is assumed that the SUs are obtained by random sampling from an infinite population, so that the error term associated with each environmental variable measured (e.g., soil texture or water content) is an independent, normally distributed variable. If this assumption does not hold, it may be possible in some cases to transform the data to meet the assumption. Alternatively, a nonparametric (distribution-free) test could be used. The advantages of using nonparametric tests is their freedom from such parametric assumptions, and they are easy to compute. They are not, however, as efficient (i.e., lack power) as the parametric tests if assumptions do hold.

In this chapter, the use of simple discriminant analysis (SDA) is presented as a useful tool for examining environmental differences among the g groups (communities). The student is most likely familiar with the univariate t-statistic and analysis of variance (AOV) tests for group differences, but is less likely to be familiar with SDA. This procedure is powerful and highly flexible in that a complete SDA generates a function that can be used (1) to determine if there is a significant difference among the g groups with regard to a number of simultaneous environmental factors (multivariate AOV); (2) to determine the multivariate distances between the g groups; (3) to determine the relative importance of each environmental factor in the separation of the g groups; and (4) to provide a maximum likelihood classification of a new SU into one of the existing g communities based on its environmental characteristics.

Historically, the application of SDA to biological data can be traced back to the simultaneous development of discriminant functions by Fisher (1936) and the D^2 multivariate distance by Mahalanobis (1936). Although the latter can be calculated from the former, the applications of discriminant functions and the Mahalanobis D^2 distance can be very different (e.g., Fisher examined the plant taxonomy of *Iris* and Mahalanobis studied anthropological dissimilarities). As originally derived by Fisher and Mahalanobis, SDA is related to multiple linear regression.

A graphical overview of SDA is given in Figure 23.1. Two groups of SUs (labeled 1 and 2) are seen plotted in the space of two environmental variables, X_1 and X_2 (the horizontal and vertical axes, respectively). In this case we assume that the two groups, each composed of seven SUs, were identified as

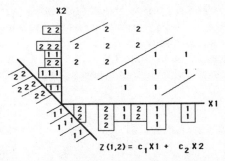

Figure 23.1   *Graphical representation of simple discriminant analysis. Two groups of SUs labeled 1 and 2, respectively, which do not overlap in the space of the environmental factors X1 and X2, are difficult to distinguish on either the X1 or X2 axis alone. The new discriminant function (Z) separates the two groups as a linear combination of X1 and X2. Modified from Legendre and Legendre (1983).*

a result of a cluster analysis based on species abundances. It can be seen from Figure 23.1 that neither X_1 nor X_2 alone distinguishes the two groups, that is, when the 14 SUs are projected onto the X_1 and X_2 axes, the SUs of groups 1 and 2 overlap along the axes. However, a discriminant function (Z), which is a linear combination of both X_1 and X_2 (i.e., a single axis Z at an angle to X_1 and X_2) separates the two groups.

23.2 PROCEDURES

Given a collection of N SUs with S species, it is assumed that $g < N$ groups have been identified using classification methods. In each SU a number of environmental variables $(X_i, i = 1, \ldots, p)$ are measured. For simplicity, the procedure of SDA outlined below has been limited to the case of $g = 2$ groups or communities (I and II), and $p = 2$ environmental variables (X_1 and X_2). The number of SUs in group I and II are N_I and N_{II}, respectively. Any number of groups (g) and environmental variables (p) could be used. But, as noted above, each group is compared to all others in a pairwise fashion in SDA and, hence, the steps described below would be repeated for each pair of groups.

STEP 1. COMPUTE MEANS, VARIANCES, AND COVARIANCES OF FACTORS. For groups I and II, the means ($\bar{x}_1$ and $\bar{x}_2$), variances (s_1^2 and s_2^2), and the covariance ($s_{1,2}$) for the two environmental variables (X_1 and X_2) are computed (see Sokal and Rohlf 1981).

STEP 2. COMPUTE THE POOLED VARIANCE–COVARIANCE MATRIX. The computation of the variance and covariance terms for the pooled matrix involves adding the sum of squares and cross product terms for the two groups and dividing by the degrees of freedom ($N_I + N_{II} - 2$). Computations are illustrated in Pielou (1984, Table 3.2) and most statistics textbooks. The pooled matrix is

$$S^* = \begin{bmatrix} s_1^{*2} & s_{1,2}^* \\ s_{2,1}^* & s_2^{*2} \end{bmatrix} \tag{23.1}$$

where the asterisk indicates a value pooled for groups I and II, s_1^{*2} and s_2^{*2} are, respectively, the variances of environmental factors X_1 and X_2, and $s_{1,2}^*$ ($= s_{2,1}^*$) is the covariance between these two variables.

STEP 3. SOLVE FOR THE COEFFICIENTS OF THE DISCRIMINANT FUNCTION (c_1 AND c_2, FIGURE 23.1). First, we set up a system of linear equations

$$c_1 s_1^{*2} + c_2 s_{1,2}^* = d_1 \tag{23.2a}$$

and

$$c_1 s_{2,1}^* + c_2 s_2^{*2} = d_2 \tag{23.2b}$$

where $d_1 = \bar{x}_{1,\text{I}} - \bar{x}_{1,\text{II}}$ and $d_2 = \bar{x}_{2,\text{I}} - \bar{x}_{2,\text{II}}$, that is, the differences in the mean vectors for groups I and II. For our simple example with two groups and two variables, coefficients c_1 and c_2 can be solved as

$$c_1 = \frac{s_2^{*2} d_1 - s_{1,2}^* d_2}{V} \tag{23.3a}$$

$$c_2 = \frac{s_1^{*2} d_2 - s_{2,1}^* d_1}{V} \tag{23.3b}$$

where $V = [(s_1^{*2})(s_2^{*2}) - (s_{1,2}^*)(s_{2,1}^*)]$, which is the determinant of Eq. (23.1).

STEP 4. *WRITE THE GENERAL DISCRIMINANT FUNCTION (Z)* between groups I and II as

$$Z_{\text{I,II}} = (c_1)(X_1) + (c_2)(X_2) \tag{23.4}$$

STEP 5. *COMPUTE THE MAHALANOBIS D^2.* The multivariate distance (D^2) between groups I and II is

$$D_{\text{I,II}}^2 = (c_1)(d_1) + (c_2)(d_2) \tag{23.5}$$

STEP 6. *TEST FOR DIFFERENCES IN GROUP CENTROIDS.* The centroids for groups I and II in the multivariate space for the two environmental variables, X_1 and X_2, are $(\bar{x}_{1,\text{I}}, \bar{x}_{2,\text{I}})$ and $(\bar{x}_{1,\text{II}}, \bar{x}_{2,\text{II}})$, respectively. The D^2 value computed in Step 5 can be used to test the hypothesis that there is no difference in these two centroids. An F statistic is computed as (Sneath and Sokal 1973)

$$F_{\text{I,II}} = \frac{D_{\text{I,II}}^2 [N_\text{I} N_\text{II}] [N_\text{I} + N_\text{II} - 3]}{2[N_\text{I} + N_\text{II}][N_\text{I} + N_\text{II} - 2]} \tag{23.6}$$

This F statistic is compared to a table of critical F values with $(2, N_\text{I} + N_\text{II} - 3)$ degrees of freedom.

STEP 7. *DETERMINE THE RELATIVE PERCENTAGE CONTRIBUTION OF X_1 AND X_2 TO D^2.* The relative percentage contributions of X_1 and X_2 to the total multivariate distance between groups I and II, denoted as RC_1 and RC_2, respectively, are computed by

$$RC_1 = \frac{c_1 d_1}{D^2} \qquad (23.7a)$$

$$RC_2 = \frac{c_2 d_2}{D^2} \qquad (23.7b)$$

STEP 8. GRAPH THE DISCRIMINANT FUNCTION (Z). Construct an axis based on the discriminant function $Z_{I,II}$ and plot the reference scores for the group centroids on this axis. The reference scores for the group centroids are obtained by substituting the means for environmental variables X_1 and X_2 into Z [Eq. (23.4)]:

$$\text{Group I:} \quad Z_I = (c_1)(\bar{x}_{1,I}) + (c_2)(\bar{x}_{2,I}) \qquad (23.8a)$$

$$\text{Group II:} \quad Z_{II} = (c_1)(\bar{x}_{1,II}) + (c_2)(\bar{x}_{2,II}) \qquad (23.8b)$$

Of course, the positions of each SU on the axis can also be calculated by substituting the values of X_1 and X_2 from each SU into $Z_{I,II}$.

23.3   EXAMPLE: CALCULATIONS

The contrived data in Table 11.3 (three species in five SUs) are expanded to include data for two environmental variables, X_1 and X_2 (Table 23.2). The five SUs have been clustered into $g = 2$ groups (I and II) based on cluster analysis (see Figure 16.2). The SDA computations on these data are illustrated below.

TABLE 23.2   *Ecological data matrix of the abundances of three species and the values for two environmental factors in five SUs. Two groups or communities (I and II) were defined from a cluster analysis*

Variables	SUs:	Group				
		I		II		
		(1)	(5)	(2)	(3)	(4)
	(1)	2	0	5	5	3
Species	(2)	0	1	3	4	2
	(3)	2	2	0	1	0
Environmental	X_1	4.2	3.8	1.6	2.4	2.0
Factors	X_2	6.1	5.9	7.7	8.3	8.0

TABLE 23.3   *Means, variances, and covariances for environmental factors X_1 and X_2 in groups I and II*

| | | Group I | | Mean | Variance | Covariance |
| | | SUs | | | | |
		(1)	(5)	$\bar{x}$	s^2	$s_{1,2}$
Environmental	X_1	4.2	3.8	4.0	0.08	0.04
Factors	X_2	6.1	5.9	6.0	0.02	

| | | Group II | | | Mean | Variance | Covariance |
| | | SUs | | | | | |
		(2)	(3)	(4)	$\bar{x}$	s^2	$s_{1,2}$
Environmental	X_1	1.6	2.4	2.0	2.0	0.16	0.12
Factors	X_2	7.7	8.3	8.0	8.0	0.09	

STEP 1.   Means, variances, and covariances are given in Table 23.3.

STEP 2.   The pooled variance–covariance matrix [Eq. (23.1)] is

$$S^* = \begin{bmatrix} 0.133 & 0.093 \\ 0.093 & 0.067 \end{bmatrix}$$

STEP 3.   Solve for the coefficients of the discriminant function. Following Eq. (23.2), we set up the system of linear equations

$$c_1(0.133) + c_2(0.093) = 4 - 2 = +2.0$$
$$c_1(0.093) + c_2(0.067) = 6 - 8 = -2.0$$

Then, from Eq. (23.3),

$$c_1 = [(0.067)(2.0) - (0.093)(-2.0)]/V = 1800$$
$$c_2 = [(0.133)(-2.0) - (0.093)(2.0)]/V = -2550$$

where $V = (0.133)(0.067) - (0.093)(0.093) = 0.000178$.

STEP 4.   Write the discriminant function, Z [Eq. (23.4)]:

$$Z = (1800)(X_1) + (-2550)(X_2)$$

STEP 5. The Mahalanobis D^2 statistic [Eq. (23.5)] is

$$D^2 = (1800)(2.0) + (-2550)(-2.0) = 8700$$

STEP 6. Test for differences in group centroids. The F test statistic is [Eq. (23.6)]

$$F = (8700)(3)(2)(3 + 2 - 3)/(2)(3 + 2)(3 + 2 - 3) = 3480$$

The probability of obtaining a D^2 of 8700 by chance alone based on an F of 3480 with df $= (2, 2)$ is less than 0.001. Thus, we reject the null hypothesis that there is no difference between the centroids of groups I and II.

STEP 7. Determine the relative percentage contribution of X_1 and X_2 to D^2. From Eq. (23.7)

$$RC_1 = (1800)(2)/8700 = 0.41 \text{ or } 41\%$$

$$RC_2 = (-2550)(2)/8700 = 0.59 \text{ or } 59\%$$

Thus, we can conclude that environmental factor X_2 is about 20% more important than factor X_1 in distinguishing between groups I and II.

STEP 8. Graph the discriminant function (Z). The reference scores for groups I and II are [Eq. (23.8)]

$$Z(\text{I}) = -8100 \quad \text{and} \quad Z(\text{II}) = -16{,}800$$

Each of the five SUs is positioned on Z by substituting their values for X_1 and X_2 into Eq. (23.4):

$$Z(1) = -7995, \quad Z(5) = -8205, \quad Z(2) = -16{,}755,$$

$$Z(3) = -16{,}845, \quad Z(4) = -16{,}800.$$

A graph of the reference scores for the means and for SUs of groups I and II along the discriminant function Z shows their separation in terms of multivariate distance (Figure 23.2).

In conclusion, the classification resulting in groups I and II was examined for possible environmental interpretation. The two variables X_1 and X_2 were found to be significantly different between the two groups and could be used to distinguish between them, with variable X_2 being somewhat more important than variable X_1.

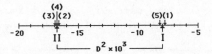

Figure 23.2   *The discriminant axis between groups I and II, with group mean locations and SU positions shown. The D^2 distance between the groups is also shown.*

TABLE 23.4   *A ecological community data matrix for five cockroach species and two environmental factors in six Panamanian localities (from Wolda et al. 1983)*

		Locality					
Species	Number	BCI (1)	LC (2)	FORT (3)	BOQ (4)	MIR (5)	CORG (6)
Ceuthobiella spp.	(1)	0	0	0	0	0	1
Compsodes cucullatus	(2)	14	38	1	1	4	0
Compsodes delicatulus	(3)	28	4	1	0	1	0
Buboblatta armata	(4)	7	0	0	0	0	0
Latindia dohrniana	(5)	68	29	0	0	11	24
Environmental Factors							
Elevation (km)	(1)	0.120	0.140	1.050	1.350	0.005	0.100
Precipitation (m)	(2)	2.5	1.5	5.0	2.5	3.0	2.5

23.4   EXAMPLE: PANAMANIAN COCKROACHES

The data set for cockroach abundances in six Panamanian localities (Table 11.4a) can be expanded to include information on two environmental factors, elevation and precipitation, provided by Wolda et al. (1983). These data are given in Table 23.4.

The classification of the cockroach data using cluster analysis (Section 16.3) suggested that the five localities could be divided into two relatively homogeneous groups: Group I defined by localities (1), (5), and (6), and Group II defined by localities (2), (3), and (4). We will use these two groups of SUs to illustrate the calculations of SDA using the BASIC program SDA.BAS. [However, note that association analysis also suggested two groups, but split locality (6) from the other five.] Although our group selection is arbitrary, we also considered ordination results in this grouping. Polar ordination positioned localities (3) and (4) close together, but apart from the other four; PCA ordination positioned localities (3), (4), and (5) near each other and separate from the spread positions of (1), (2), and (6).

TABLE 23.5   *Discriminant analysis results from SDA.BAS using elevation* (X_1) *and precipitation* (X_2) *for the interpretation of differences between two cockroach locality groups*

Mean, variance (Var), standard deviation (SD), standard error (SE), and covariance (Covar) statistics for environmental factors within groups:

Group	Factor	Mean	Var	SD	SE	Covar
I	X_1	0.075	0.004	0.061	0.035	-0.018
	X_2	2.667	0.083	0.289	0.167	
II	X_1	0.847	0.397	0.630	0.364	0.607
	X_2	3.000	3.250	1.803	1.041	

Discriminant coefficients: $c_1 = -4.809$ and $c_2 = 0.651$
Mahalanobis distance = 3.494
F statistic = 1.97 at df = (2, 3)
Contribution of each environmental factor

Environmental Factor	Relative% Contribution
X_1	106.2
X_2	-6.2

Group centroids on the discriminant axis:

$$Z_I = 1.38 \quad \text{and} \quad Z_{II} = -2.12$$

SU scores along the discriminant axis

Group:	I			II		
SU:	(1)	(5)	(6)	(2)	(3)	(4)
Score:	1.05	1.93	1.15	0.30	-1.79	-4.86

The results of the SDA on these data indicate a poor discrimination between these two cockroach groups based on elevation and precipitation (Table 23.5). In fact, the results illustrate an interesting property of SDA in that the inclusion of the second factor (precipitation) actually weakened the analysis, since its relative percentage contribution to multivariate distance was actually negative (-6.2%). Thus, the inclusion of a weak discriminating variable into an SDA can actually lessen the strength of the analysis; this is not true with multiple linear regression (Sokal and Rohlf 1981), where the addition of variables always increases the strength of the analysis (albeit often insignificantly).

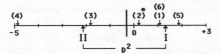

Figure 23.3   *Simple discriminant analysis of six Panamanian cockroach localities showing the reference scores for groups I and II and the six localities along the discriminate function. The Mahalanobis D^2 distance is also shown. *Locality (2) is misclassified.*

TABLE 23.6   *Ecological data matrix for two site factors in 10 upland forest sites, southern Wisconsin. For details on these factors and the corresponding tree data see Peet and Loucks (1977) and Gauch (1982)*

Site Factor	Number	Forest Sites (SUs)									
		(1)	(2)	(3)	(4)	(5)	(6)	(7)	(8)	(9)	(10)
Soil texture	(1)	4	5	3	2	1	1	2	1	1	1
Stand dynamics	(2)	1	2	1	2	1	3	3	5	4	5

The position of the two cockroach groups (I and II) along the linear discriminant function is shown in Figure 23.3, along with the positions for each of the six localities. The spread of the three localities [(2), (3), and (4)] about the reference score for group II further substantiates the poor discrimination using elevation and precipitation as factors. Locality (2) is actually "misclassified" into group I using the halfway point between the group reference scores as the dividing line for *minimum likelihood* misclassification (Sneath and Sokal 1973).

23.5   EXAMPLE: WISCONSIN FORESTS

We expanded the data matrix for eight trees in 10 upland forest sites (Table 11.6a) to include information in Peet and Loucks (1977) on site factors (Table 23.6). They provided information for soil texture (five classes of percentage sand in the A1 horizon), which they viewed as an index defining a moisture–nutrient regime. They also provided an index of stand dynamics, which can be viewed as an index of species turnover rates; this index is related to site quality.

Results of the cluster analysis (Figure 16.4) were used to place the 10 SUs into two groups: Group I, defined by sites (1) to (4), and Group II defined by sites (5) to (10). The results suggest a significant multivariate distance

TABLE 23.7   *An SDA between two upland forest communities or groups of SUs,*
southern Wisconsin, using two factors for discrimination: X_1 = soil texture (five classes
of percentage sand), and X_2 = index of site dynamics as indicated by species turnover
rate (see Peet and Loucks 1977)

Mean, variance (Var), standard deviation (SD), standard error (SE), and covariance
(Covar) statistics for environmental factors within groups:

Group	Factor	Mean	Var	SD	SE	Covar
I	X_1	3.50	1.67	1.29	0.645	<0.001
	X_2	1.50	0.33	0.58	0.289	
II	X_1	1.17	0.17	0.41	0.167	−0.100
	X_2	3.50	2.30	1.52	0.619	

Discriminant coefficients: c_1 = 3.10 and c_2 = −1.16
Mahalanobis distance = 9.547
F statistic = 10.0 at df = (2, 7)
Contribution

Environmental Factor	Relative% Contribution
X_1	75.8
X_2	24.2

Group centroids on the discriminant axis:

$$Z_I = 9.1 \quad \text{and} \quad Z_{II} = -0.4$$

SU scores along the discriminant axis

Group:	I				II					
SU:	(1)	(2)	(3)	(4)	(5)	(7)	(6)	(8)	(9)	(10)
Score:	11.2	13.2	8.1	3.9	1.9	2.7	−0.4	−2.7	−1.5	−2.7

(D^2 = 9.5) between these two groups using soil texture and stand dynamics
(Table 23.7). The probability is less than 1% that the two group centroids are
equal (F = 10.0). Soil texture is about three times as important in discriminating the two groups as the stand dynamics index.

The positions of the group reference scores (based on their centroids) and
the 10 sites (SUs) along the discriminant function (Figure 23.4) illustrate the
separation of the two groups. Note that the sites are not tightly clustered about
the group I reference score and, in fact, site (4) is misclassified into group II
based on the maximum likelihood criterion.

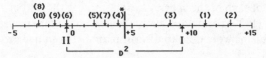

Figure 23.4 *Simple discriminant analysis of 10 Wisconsin forest sites showing the reference scores for groups I and II and the 10 sites along the discriminant function. The Mahalanobis D^2 distance is also shown. *Site (4) is misclassified.*

23.6   ADDITIONAL TOPICS IN CLASSIFICATION INTERPRETATION

SDA compares each group or community to others in a pairwise fashion, that is, a discriminant function is determined for each pair of communities being contrasted, a D^2 statistic is computed to measured the multivariate distance between each pair, and an F ratio is used to test the statistical significance of this distance. However, interest may center on the comparison of three or more communities taken *simultaneously* (e.g., Gerdol et al. 1985, Walker et al. 1979, and Wiegleb 1981). In this case, the discriminant function, the multivariate distance, and the F ratio define overall differences between all the groups taken together. This method is commonly called *multiple discriminant analysis* (MDA) or *canonical variate analysis*.

MDA involves the computation of eigenvalues and vectors on pooled variance–covariance matrices and is not illustrated in this chapter. However, the student will find the transition to a simultaneous comparison of three or more groups (MDA) relatively easy after obtaining a feeling for the simpler SDA procedures presented here. With SDA (two groups) a single discriminant function axis is derived, but with MDA (three or more groups) two or more discriminant function (canonical) axes are derived. The student is referred to treatments of MDA by Gittens (1985), Legendre and Legendre (1983), Orloci and Kenkel (1985), Pielou (1984), and Williams (1983).

In Chapter 22, we stated that classification may be considered a two-step process: (1) a reduction of N SUs into $g < N$ groups based on species abundances and (2) testing for statistical differences in environmental factors between the g groups. This seems a logical process where communities are first defined by biota and then examined for abiotic differences, which might help explain the biotic responses. While the second step was emphasized in this chapter, many of the univariate and multivariate statistical tests are also appropriate for examining the importance of species in determining the g groups. By analogy, all of the questions posed with regards to the abiotic factors could be also done for the S species. For example, is there a significant difference in the species centroids of the g groups? What is the relative contribution of each of the species in determining the D^2 between the groups? Walker et al. (1979) conducted such a study where the species abundances of benthic organisms were used to discriminate between five groups of substrate classes.

The environmental interpretation of ecological classifications involves a great deal more than familiarity with multivariate procedures like SDA and MDA. The ecologist must wrestle with the ubiquitous (and often ambiguous) problem of what environmental factors to include in the multivariate analysis. Should the analysis only include those variables that have a potential (in the ecologist's view) of being significant, perhaps basing this judgment on some preliminary univariate tests? Or, should the analysis include only those variables which are *operational* (Spomer 1973), that is, have a potential for a direct interaction (involving energy or mass) with the organisms (e.g., water and nutrients)? Alternatively, is it adequate to include more easily measured *nonoperational* variables (e.g., site elevation, aspect slope) that indirectly involve operational variables?

By including a large number of variables (which most likely will include some "weak" ones), the discriminant analysis might actually be weakened (see the example in Section 23.4). Also, using a large number of variables introduces the risk of including highly correlated variables, which often causes problems with multivariate statistical procedures (Sokal and Rohlf 1981). For example, Eq. 23.7) assumes uncorrelated variables (Sneath and Sokal 1973). Interactions between ecological variables are, of course, the expected, rather than the exception, and one of the rationales for using multivariate statistical tests over the simpler univariate tests is to obtain a better interpretation of where interactions exist (i.e., as the covariances are used in a SDA).

The ecologist must take great care in selection (and measurement) of all ecological factors to be included in a multivariate analysis. For this reason we do not recommend the use of *stepwise* SDA or MDA, where the tendency is to "throw" all measured variables (important or not) into an MDA and let the stepwise procedure select those "statistically" important variables; many of these variables may have little ecological importance. The stepwise procedure will often mix operational with nonoperational factors, potentially leading to a more difficult ecological interpretation. Given that the ecologist uses good judgment in the selection of environmental factors, SDA and MDA can be powerful tools for interpreting the patterns of community variation.

23.7   SUMMARY AND RECOMMENDATIONS

1. When interesting patterns in SU clusters are found from a classification, the next logical step is to provide an environmental interpretation for this

pattern. Simple discriminant analysis (SDA) is a powerful multivariate statistical method if there are significant differences in environmental factors (taken together) between pairs of community clusters (groups of SUs). SDA also provides information on the relative importance of each environmental factor in the discrimination (Section 23.1).

2. For the ecologist who is interested in comparing three or more community groups simultaneously (rather than in pairs as with SDA), we recommend the use of multiple discriminant analysis (MDA; also called *canonical variate analysis*).

3. For the simple purpose of illustrating the procedures and computations for a discriminant analysis, we used small data sets. However, when students apply such methods to their own data, we recommend that careful attention be paid to the underlying parametric and multivariate assumptions. Sample size, particularly the number of SUs in each group, is a very important consideration. If the sample size is small and, therefore, the underlying assumptions are unlikely to hold, we recommend the use of nonparametric multivariate methods.

4. With SDA, the inclusion of "weak" discriminating variables in the analysis can decrease the power of the analysis (see, e.g., Section 23.4). Thus, we recommend great care be taken in selecting environmental factors to include in an SDA or MDA. Those environmental factors of high ecological importance, rather than purely statistical importance, should be used. We do not recommend the use of stepwise SDA or MDA (Section 23.6).

CHAPTER 24

Ordination Interpretation

In this chapter we present various strategies for relating SU ordination coordinates (derived from species data) to environmental factors. These ordination coordinates may be conceptualized as synthetic axes that bear some relationship to underlying environmental gradients. We illustrate the application of simple and multiple linear regression for examining these relationships.

24.1 GENERAL APPROACH

A central theme in plant ecology is the elucidation of the relationships between vegetation and environmental factors. This has given rise to a suite of techniques known as *direct* and *indirect gradient analysis* (defined in Chapter 17). Strategies for interpreting indirect ordinations vary widely. Bray and Curtis (1957) used simple graphs of species importance and measures of environmental factors within SUs in a polar ordination (PO). Some have tried a principal component analysis (PCA) on mixed data matrices of species and environmental factors (Barkham and Norris 1970, Walker and Wehrhahn 1971). A variety of mathematical techniques have been used, including canonical correlation analysis (Gauch and Wentworth 1976), correspondence analysis (COA), and residual ordination analysis (Carleton 1984).

A general model developed to address questions of environmental interpretation of ordinations is illustrated in Figure 24.1. Recall that the community data matrix has two subsets of variables, one for species and the other for environmental factors (e.g., Figure 22.1). Given these two subsets, three strategies are shown, all of which may involve the use of correlation and

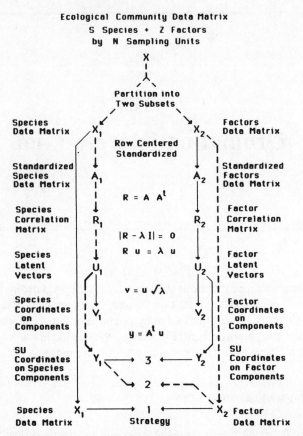

Figure 24.1 *Strategies for relating species abundances or synthetic species components (from PCA, COA, or DPC) with environmental factors (direct gradients) or environmental components (indirect synthetic gradients). The scheme illustrated in this chapter is Strategy 2 (dashed lines). Matrix notation follows that of Chapter 19.*

regression analysis. The most direct approach (Strategy 1) is to relate the abundance of each species to each environmental factor across the N SUs by correlation or regression. In Strategy 2, we take the SU coordinates derived from an eigenanalysis of species data (e.g., PCA, COA) and regress these coordinates onto the environmental factors measured in these SUs. It is also possible to derive SU ordination coordinates from an eigenanalysis of both species and factors data and correlate these results (Strategy 3).

Strategy 1 is an example of a direct gradient analysis, whereas strategies 2 and 3 are techniques of indirect gradient analysis (Austin et al. 1984, Gauch

1982, Whittaker 1967). In this chapter, we illustrate Strategy 2 (shown as the dashed line in Figure 24.1).

24.2   PROCEDURES

The following steps illustrate Strategy 2:

STEP 1. OBTAIN SU COORDINATES.   SU coordinates on ordination axes of PCA or COA (or DPC) are obtained. Note, from Figure 24.1, that SU coordinates are obtained from a species correlation matrix R. (If results from a DPC suggest a significant nonlinear structure to the data, then the "unfolded" DPC axis is used rather than (or along with) the first two PCA axes.)

STEP 2. REGRESSION ANALYSIS.   The simple and multiple linear regression coefficients that relate SU coordinates to individual environmental variables in each respective SU are computed.

a. The simple linear regression (SLR) model is

$$Y = B_0 + B_1 X \tag{24.1}$$

where Y represents ordination coordinates for the SUs and X represents values for the particular environmental factor of interest that has been measured in the SUs. Equation (24.1) is solved using least-squares techniques (see Box 14.1, Sokal and Rohlf 1981). First, we calculate means, $\bar{y}$ and $\bar{x}$, corrected sum of squares, Σy^2 and Σx^2, for Y and X, and corrected sum of cross products, Σxy, between X and Y. Then, coefficient B_1 (the slope) is given by

$$B_1 = \frac{\Sigma xy}{\Sigma x^2} \tag{24.2}$$

and B_0 (the intercept) is

$$B_0 = \bar{y} - (B_1 \bar{x}) \tag{24.3}$$

Also of interest is the coefficient of determination, r^2, an estimate of the amount of variation in Y accounted for by the regression of Y onto X:

$$r^2 = \frac{(\Sigma xy)^2/\Sigma x^2}{\Sigma y^2} \tag{24.4}$$

The statistical significance of the SLR is determined by computing the square root of r^2 (i.e., r) and comparing it to a table of probabilities for r at $P = 0.05$ with $N - 2$ degrees of freedom (e.g., Table A.13, Steele and Torrie 1960) or by determining a t-statistic for B_1 (see Box 14.3, Sokal and Rohlf 1981).

  b. The multiple linear regression (MLR) model is

$$Y = b_0 + (b_1 X_1) + (b_2 X_2) + \cdots + (b_p X_p) \qquad (24.5)$$

where Y is defined as above [Eq. (24.1)] and the X's represent the values for the p environmental factors that have been measured in the SUs. Again, we calculate means for Y (i.e., $\bar{y}$) and for each X (i.e., $\bar{x}_1, \bar{x}_2, \cdots \bar{x}_p$) the corrected sum of squares for Y (i.e., Σy^2) and each X (e.g., Σx_1^2) and the corrected sum of cross products between each X and Y (e.g., $\Sigma x_2 y$), and between the X's (e.g., $\Sigma x_1 x_2$). Then, the partial regression coefficients (b's) are determined by solving a system of simultaneous linear (normal) equations (Sokal and Rohlf 1981). These normal equations are readily solved using computerized algorithms. If we take the simple case of two equations in two unknowns (two environmental factors), the equations are

$$b_1 \Sigma x_1^2 + b_2 \Sigma x_1 x_2 = \Sigma x_1 y \qquad (24.6a)$$

$$b_1 \Sigma x_2 x_1 + b_2 \Sigma x_2^2 = \Sigma x_2 y \qquad (24.6b)$$

The coefficients b_1 and b_2 are solved using

$$b_1 = \frac{(\Sigma x_2^2 \Sigma x_1 y) - (\Sigma x_1 x_2 \Sigma x_2 y)}{V} \qquad (24.7a)$$

$$b_2 = \frac{(\Sigma x_1^2 \Sigma x_2 y) - (\Sigma x_2 x_1 \Sigma x_1 y)}{V} \qquad (24.7b)$$

where $V = (\Sigma x_1^2 \Sigma x_2^2) - (\Sigma x_1 x_2 \Sigma x_2 x_1)$, and

$$b_0 = \bar{y} - [(b_1 \bar{x}_1) + (b_2 \bar{x}_2)] \qquad (24.7c)$$

The coefficient of multiple determination, R^2, is

$$R^2 = \frac{(b_1 \Sigma x_1 y) + (b_2 \Sigma x_2 y)}{\Sigma y^2} \qquad (24.8)$$

which is an estimate of the proportion of the variation in Y accounted for jointly by the X's. The statistical significance of MLR is estimated by determining an F-ratio (e.g., see Box 16.2, Sokal and Rohlf 1981).

STEP 3. EXAMINE RESULTS. We interpret the results by examining the direction (sign) and statistical significance of the regression coefficients. This will indicate which ordination components are related to which environmental factors (gradients), if any.

It must be emphasized that even strong statistical correlations and regressions may or may not represent meaningful ecological relationships; it always behooves the ecologist to closely scrutinize all results and, perhaps, conduct additional studies where statistical results suggest possible relationships. Also remember that ordination components are synthetic (in that they are derived by such methods as PCA or DPC) and may or may not correspond to meaningful underlying ecological trends.

24.3 EXAMPLE: CALCULATIONS

We illustrate the calculations for an environmental interpretation of ordinations by using the contrived data for two species and two environmental factors in five SUs (Table 23.2). From these data, SU scores on components I and II from PCA (Figure 19.4) and on the detrended principal component (Figure 21.2) will be used, along with the values for environmental factors, X_1 and X_2. This simplified data set is sufficient to illustrate the computations for Strategy 2 (Figure 24.1).

STEP 1. OBTAIN THE SU COORDINATES. These results are summarized in Table 24.1.

STEP 2. REGRESSION ANALYSIS. Coefficients for simple and multiple regression of ordination coordinates onto the environmental factors can now be obtained. To illustrate the computations, we have selected DPC(I) coordinates (Table 24.1) as the dependent variable, Y, in Eqs. (24.1) and (24.5).

a. Simple linear regression. Given $Y = Y_3$ [i.e., DPC(I)] and $X = X_1$ (Table 24.1), the slope and intercept coefficients are computed following Eqs. (24.2) and (24.3):

$$B_1 = -3.064/5.2 = -0.589$$
$$B_0 = 0.026 - (-0.589)(2.8) = 1.675$$

TABLE 24.1   *Coordinates for five SUs on principal components (PC) I and II and on the detrended principal component (DPC) along with the measures for two environmental factors, X_1 and X_2, in the SUs. Means, corrected sum of squares, and corrected sum of cross products are given for the components and factors*

SUs	Components				Environmental Factors	
	PC(I) Y_1	PC(II) Y_2	DPC(I) Y_3		X_1	X_2
(1)	−0.62	−0.28	−0.58		4.2	6.1
(2)	+0.56	−0.12	+0.71		1.6	7.7
(3)	+0.78	+0.12	+1.00		2.4	8.3
(4)	0.00	−0.50	0.00		2.0	8.0
(5)	−0.72	+0.28	−1.00		3.8	5.9
$\bar{y} =$	0.00	−0.10	0.026	$\bar{x} =$	2.8	7.2
$\Sigma y^2 =$	1.825	0.386	2.837	$\Sigma x^2 =$	5.2	5.0
$\Sigma x_1 y =$	−2.572	0.384	−3.064	$\Sigma x_1 x_2 =$	−4.520	
$\Sigma x_2 y =$	2.756	−0.384	3.393	$\Sigma x_2 x_1 = \Sigma x_1 x_2$		

with a coefficient of determination [Eq. (24.4)]

$$r^2 = [(-3.064)(-3.064)/(5.2)]/2.837 = 0.636$$

which, given $r = 0.798$, is not significant at $P = 0.05$ (df = 3).
For environmental factor X_2,

$$B_1 = 3.393/5.0 = 0.679$$
$$B_0 = 0.026 - (0.679)(7.2) = -4.862$$

with a coefficient of determination

$$r^2 = [(3.393)(3.393)/5.0)]/2.837 = 0.812$$

which, given $r = 0.901$, is significant at $P = 0.05$ (df = 3).

In conclusion, the DPC scores are significantly related to environmental factor X_2. The SUs are positioned along the detrended principal component based on overall variations in their species abundances and these variations are significantly related to an underlying X_2 gradient. A plot of Y [DPC(I) scores] against X_2 illustrates this relationship (Figure 24.2).

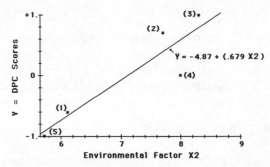

Figure 24.2 *The linear regression relationship between SU scores on the detrended principal component and the environmental factor X2.*

b. Multiple linear regression. The slope and intercept coefficients for $Y =$ DPC(I) scores are determined by first solving Eq. (24.6):

$$(b_1)(5.2) + (b_2)(-4.520) = -3.064$$
$$(b_1)(-4.520) + (b_2)(5.0) = 3.393$$

and, then, from Eq. (24.7),

$$b_1 = [(5.0)(-3.064) - (-4.520)(3.393)]/5.570 = +0.0029$$
$$b_2 = [(5.2)(3.393) - (-4.520)(-3.064)]/5.570 = 0.681$$

given $V = (5.2)(5.0) - (-4.520)(-4.520) = 5.570$

$$b_0 = 0.026 - [(-0.0029)(2.8) + (0.681)(7.2)] = -4.869$$

The coefficient of multiple determination [Eq. (24.8)]

$$R^2 = [(-0.0029)(-3.064) + (0.681)(3.393)]/2.837 = 0.818$$

which, given $R = 0.904$, is not significant at $P = 0.05$, because of low degrees of freedom (df $= 2$).

Even though the MLR is not statistically significant in this simple data set, we will illustrate how the relative percentage contribution of each factor to the total coefficient of determination (R^2) can be determined, using analogous calculations to those for determining the relative contribution of factors to the Mahalanobis distance (D^2) in SDA (Chapter 23). The relative contributions of factors X_1 and X_2 are

$$RC_1 = [(-0.0029)(-3.064)/(2.837)/0.818](100\%) = 0.4\%$$

$$RC_2 = [(0.681)(3.393)/(2.837)/0.818](100\%) = 99.6\%$$

It can be seen that environmental factor X_2 is far more important than factor X_1 for explaining the variation in SU positions along DPC(I).

24.4 EXAMPLE: PANAMANIAN COCKROACHES

In Table 23.4 the data set for the six Panamanian cockroach localities includes a subset of two environmental factors: elevation and precipitation. Using these

TABLE 24.2 *Simple and multiple linear regression of component scores for six Panamanian localities onto two environmental factors, elevation (X_1) and precipitation (X_2)*

Statistics for environmental factors and principal components

Variable	Mean	Variance	Standard Deviation	Standard Error
Factor X_1	0.461	0.339	0.582	0.238
Factor X_2	2.833	1.367	1.169	0.477
Component I	−0.000	0.605	0.778	0.318
Component II	−0.000	0.235	0.485	0.198

Simple linear regression

Comparison PC Factor		Regression Coefficient B_0	Coefficient B_1	r^2	Standard Error of B_1	t Statistic	df
I	X_1	−0.261	0.566	0.180	0.605	0.936	4
I	X_2	−0.611	0.216	0.105	0.315	0.685	4
II	X_1	0.071	−0.154	0.034	0.409	−0.377	4
II	X_2	−0.131	0.046	0.012	0.206	0.225	4

Multiple linear regression

Component I: $Y = -0.523 + (0.466X_1) + (0.109X_2)$

Component II: $Y = -0.178 + (-0.250X_1) + (0.104X_2)$

Coefficients of multiple determination, F statistics, and df

	R^2	F	df
Component I:	0.201	0.388	2, 3
Component II:	0.083	0.136	2, 3

data, along with the results of a PCA (Table 19.4), SLR and MLR analyses were run using the BASIC program PCREG.BAS (accompanying disk). [Because the results for DPC analysis on this data set were not significant (Table 21.2), we have not included a detrended component.] Output from program PCREG.BAS for the regression of principal components I and II onto factors X_1 (elevation) and X_2 (precipitation) suggest no significant linear relationships between components and factors (Table 24.2). The r^2 (SLR) and R^2 (MLR) were both low and insignificant.

24.5   EXAMPLE: WISCONSIN FORESTS

In Table 23.6 two factors, soil texture and an index of "site dynamics," are shown, corresponding to the species data for the 10 upland Wisconsin forest sites. Using the results for PCA (Table 19.5) and DPC (Figure 21.4), results for SLR and MLR analyses using PCREG.BAS are given in Table 24.3. Note that PC(I) and DPC are negatively related to factor X_2 (index of site dynamics) and positively related to factor X_1 (soil texture).

The coefficients of determination are high and the t and F statistics are highly significant ($P < 0.001$). PC(II) was not significantly related to the factors by either SLR or MLR, as expected, because it represents an "arch," effectively removed by the DPC analysis.

Graphs of SU coordinates on PC(I) and DPC against values for factor X_1 (Figure 24.3a) and factor X_2 (Figure 24.3b) clearly show these linear relationships.

24.6   ADDITIONAL TOPICS IN ORDINATION INTERPRETATION

In this chapter we presented strategy 2 (Figure 24.1), for regressing components (coordinates) derived from an ordination (e.g., PCA and DPC) directly onto *abiotic* (environmental) factors taken either one at a time (SLR) or together (MLR). Species-abundance data are used to generate those components for the SUs, which are then examined with regard to the environmental factor data subset. We now briefly describe other strategies where species and environmental data are simultaneously analyzed to elucidate relationships.

At first it seems intuitive to treat the biotic and abiotic subsets of the ecological community data matrix as one large data set and subject this matrix to, for example, a PCA and examine the correspondence (relative positions) of SU coordinates for both species and environmental factors. However, components extracted by a PCA on this type of mixed data are no longer biotic components but, rather, are confounded biotic–abiotic

TABLE 24.3 *Results for simple and multiple linear regression of SU coordinates from principal components, PC(I) and PC(II), and detrended principal components DPC(III) onto two factors, soil texture (X_1) and an index of site dynamics (X_2)*

Statistics for environmental factors and principal components

Variable	Mean	Variance	Standard Deviation	Standard Error
Factor X_1	2.100	2.100	1.449	0.458
Factor X_2	2.700	2.456	1.567	0.496
Component PC(I)	0.000	0.492	0.702	0.222
Component PC(II)	0.000	0.131	0.363	0.115
Component DPC	0.074	0.620	0.788	0.249

Simple linear regression

Comparison Variables	Regression Coefficients B_0	Coefficients B_1	r^2	Standard Error of B_1	t Statistic	df
PC(I), X_1	−0.729	0.347	0.514	0.119	2.907	8
PC(I), X_2	1.082	−0.401	0.801	0.070	−5.675	8
PC(II), X_1	−0.283	0.135	0.291	0.074	1.813	8
PC(II), X_2	−0.058	0.021	0.008	0.081	0.266	8
DPC, X_1	−0.818	0.425	0.610	0.120	3.540	8
DPC, X_2	1.275	−0.445	0.783	0.082	−5.369	8

Multiple linear regression

Variable	Multiple Regression Coefficients b_0	b_1	b_2	R^2	F	df
PC(I)	0.563	0.147	−0.323	0.863	22.07	2, 7
PC(II)	−0.838	0.221	0.139	0.532	3.98	2, 7
DPC	0.491	0.222	−0.327	0.895	29.86	2, 7

components. Ecologically, the objective is to extract biotic components and interpret them with regard to abiotic influences. This becomes difficult, if not impossible, with confounded biotic–abiotic components, since the results are strongly influenced by correlations within species, within factors, and between species and factors. In fact, since species and environmental factors will not be measured in the same units, these data must be standardized in some fashion.

Walker and Wehrhahn (1971) described a procedure performing a PCA on a combined species-factor data set such that the extracted components reflect

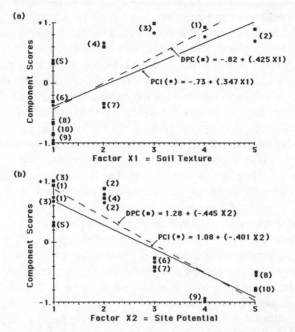

Figure 24.3 *Plot of scores for 10 upland Wisconsin forest sites on the first principal component (PC I) and the detrended component (DPC) against (a) factor X1— soil texture, and (b) factor X2—site dynamics index.*

variations in the biotic subset and the abiotic subset does not unduly influence the nature of the components. Basically, their procedure involves standardizing the abiotic data by scaling each factor from 0 to 1 by dividing by the maximum value for each factor and then scaling these down further to very small values (by dividing by 1000) before entering these into a PCA. The result is that the total variance–covariance contribution of these scaled abiotic factors will be small relative to that for the biotic variables and, thus, the extracted components will largely reflect biotic components of variation. The coordinates for abiotic factors can subsequently be adjusted to provide their correlations with the biotic components. Walker and Wehrhahn (1971) obtained ecologically interpretable results using this procedure on wetlands data in Canada.

Canonical correlation analysis (Gauch and Wentworth 1976) is another procedure for simultaneously analyzing mixed species and environmental data. This multivariate statistical procedure uses linear canonical equations with multiple dependent and independent variables (in contrast to multiple linear regression that only uses multiple independent variables). The *partial* canonical coefficients are determined such as to maximize the correlation between the dependent variable (species) subset and the independent variable (environmental factors) subset. Intuitively, this procedure seems ideal for ecological interpretation as the species and factor coordinates can be plotted in the same *canonical* ordination system and their relative positions define their interrelationships. However, many applications of canonical correlation analysis on ecological data have produced results of limited usefulness (e.g., Austin 1968). The assumption of linearity for both the species-abundance relations and the factor gradients appears to be quite stringent (Gauch and Wentworth 1976). Of course, the linear regression approaches presented in this chapter have similar assumptions; nonlinear regression is discussed below.

Canonical correspondence analysis is a new procedure that combines within one algorithm a reciprocal averaging solution for a correspondence analysis on species-site data (with a detrending option) and a weighted multiple regression analysis on environmental factor-site data (Ter Braak 1986, 1987). A comparison of this canonical correspondence analysis (detrended and undetrended) with detrended correspondence analysis on three data sets indicates that detrended canonical correspondence analysis very effectively provided ecological interpretations (Ter Braak 1986). Although the details of this algorithm are beyond the scope of this chapter, we recommend the student be aware of further applications and testing of this method. A four-stage procedure that uses detrended correspondence analysis, rank correlations, and nonmetric multidimensional scaling to successfully identify vegetation-environmental factor complex has been described by Whittaker (1987). Another new procedure is residual ordination analysis, which puts emphasis on explaining the residual variation about ordination components (Carleton 1984).

Although the treatments in this chapter are limited to simple linear and multiple linear regressions, various curvilinear regression procedures could also be used to relate a component [e.g., PC(I)] to a single environmental factor (e.g., soil moisture). A number of such curvilinear regression models are possible from relatively simple exponential and power functions (e.g., Legendre and Legendre 1983) to relatively complex Gaussian models (e.g., Gauch 1982). For relating a component to many factors taken together, multiple curvilinear regression models could also be used (e.g., polynomial regressions—Sokal and Rohlf 1981).

24.7 SUMMARY AND RECOMMENDATIONS

1. When searching for an environmental interpretation to an interesting patterning of SUs along an ordination axis, the strategy of using linear

regression to relate SU coordinates on the axis to environmental variables can provide meaningful interpretations (Section 24.1), providing the gradient is narrow and the species responses are linear (Figure 21.1a).

2. Both simple and multiple linear regression can be used to explore the relationships between ordination axes and environmental factors. While we recommend both, we suggest that multiple regression is likely to be more informative, since ecological community data are multivariate in nature. The student is also reminded that statistically significant regressions may or may not imply meaningful ecological relationships (Section 24.2).

3. We do not recommend the use of a principal components analysis on one large data set of combined species and environmental variables. The components extracted by this PCA will be confounded biotic–abiotic components, making the ecological interpretation of species variations on site factors unclear. Although the site factors can be scaled down to very small values, hence contributing little to extracted principal components, the use of methods specifically designed to simultaneously analyze species–environmental relationships are preferred (Section 24.6).

4. Canonical correspondence analysis is a new procedure that appears to have promise for the simultaneous exploration of site species-factor data. It combines into one algorithm detrended correspondence analysis on species data with weighted multiple regression on environmental data. Further developments of this technique may result in an ecologically powerful tool.

References

Abrams, P. A. (1980). Some comments on measuring niche overlap. *Ecology* **61**: 44–49.

Alatalo, R. V. (1981). Problems in the measurement of evenness in ecology. *Oikos* **37**: 199–204.

Alatalo, R. V. and R. Alatalo (1977). Components of diversity: multivariate analysis with interaction. *Ecology* **58**: 900–906.

Able, K. P. and B. R. Noon (1976). Avian community structure along elevational gradients in the northeastern United States. *Oecologia* **26**: 275–294.

Anderberg, M. R. (1973). *Cluster Analysis for Applications.* Academic Press, New York.

Anderson, A. J. B. (1971). Ordination methods in ecology. *Journal of Ecology* **59**: 713–726.

Anderson, D. J. (1965). Classification and ordination in vegetation science: controversy over a nonexistent problem? *Journal of Ecology* **53**: 521–526.

Andrews, M. J. and D. Richard (1980). Rehabilitation of the Inner Thames estuary. *Marine Pollution Bulletin* **11**: 327–331.

Archie, J. W. (1984). A new look at the predictive value of numerical classification. *Systematic Zoology* **33**: 30–51.

Ashton, D. H. (1976). The vegetation of Mount Piper, Central Victoria: A study of a continuum. *Journal of Ecology* **64**: 463–483.

Austin, M. P. (1968). An ordination study of a chalk grassland community. *Journal of Ecology* **56**: 739–757.

Austin, M. P. (1985). Continuum concept, ordination methods and niche theory. *Annual Review of Ecology and Systematics* **16**: 39–61.

Austin, M. P. and L. Orloci (1966). Geometric models in ecology. II. An evaluation of some ordination methods. *Journal of Ecology* **55**: 217–227.

Austin, M. P., R. B. Cunningham, and P. M. Fleming (1984). New approaches to direct gradient analysis using environmental scaling and statistical curve-fitting procedures. *Vegetatio* 55:11–28.

Bannister, P. (1968). An evaluation of some procedures used in simple ordinations. *Journal of Ecology* 54:665–674.

Barkham, J. P. and J. M. Norris (1970). Multivariate procedures in an investigation of vegetation and soil relations of two beech woodlands, Cotswold Hills, England. *Ecology* 51:630–639.

Bartlett, M. S. (1954). A note on the multiplying factors for various chi-squared approximations. *Journal Royal Statistical Society, Series B* 16:296–298.

Beals, E. W. (1960). Forest bird communities in the Apostle Islands of Wisconsin. *Wilson Bulletin* 72:156–181.

Beals, E. W. (1965). Ordination of some corticolous cryptogamic communities of south-central Wisconsin. *Oikos* 16:1–8.

Beals, E. W. (1984). Bray–Curtis ordination: An effective strategy for analysis of multivariate ecological data. *Advances in Ecological Research* 14:1–55.

Bennett, S. H., J. Whitfield Gibbons, and J. Glanville (1980). Terrestrial activity, abundance and diversity of amphibians in differently managed forest types. *American Midland Naturalist* 103:412–416.

Bliss, C. I. and D. W. Calhoun (1954). *An Outline of Biometry.* Yale Co-op Publ., New Haven, CT.

Bliss, C. I. and R. A. Fisher (1953). Fitting the binomial distribution to biological data and a note on the efficient fitting of the negative binomial. *Biometrics* 9:176–200.

Bloom, S. A. (1981). Similarity indices in community studies: Potential pitfalls. *Marine Ecology* 5:125–128.

Boesch, D. F. (1977). *Application of Numerical Classification in Ecological Investigations of Water Pollution.* EPA-600/3-77-033. U.S. Environmental Protection Agency, Corvallis, OR.

Bradfield, G. E. and N. C. Kenkel (1987). Nonlinear ordination using flexible shortest path adjustment of ecological distances. *Ecology* 68:750–753.

Bray, J. R. and J. T. Curtis (1957). An ordination of the upland forest communities of southern Wisconsin. *Ecological Monographs* 27:325–349.

Brown, M. J., D. A. Ratkowsky and P. R. Minchin (1984). A comparison of detrended correspondence analysis and principal co-ordinates analysis using four sets of Tasmanian vegetation data. *Australian Journal of Ecology* 9:273–279.

Brown, R. T. and J. T. Curtis (1952). The upland conifer-hardwood forests of northern Wisconsin. *Ecological Monographs* 22:217–234.

Bultman, T. L. and S. H. Faeth (1985). Patterns of intra- and interspecific associations in leaf-mining insects on three oak host species. *Ecological Entomology* 10:121–129.

Campbell, B. M. (1978). Similarity coefficients for classifying releves. *Vegetatio* 37:101–109.

Carleton, T. J. (1980). Non-centered component analysis of vegetation data: a comparison of orthogonal and oblique rotation. *Vegetatio* **42**:59–66.

Carleton, T. J. (1984). Residual ordination analysis: a method for exploring vegetation-environment relationships. *Ecology* **65**:469–477.

Carpenter, S. R. and J. E. Chaney (1983). Scale of spatial pattern: four methods compared. *Vegetatio* **53**:153–160.

Cattell, R. B. (1952). *Factor Analysis*. Harper, New York.

Chang, D. H. S. and H. G. Gauch, Jr. (1986). Multivariate analysis of plant communities and environmental factors in Ngari, Tibet. *Ecology* **67**:1568–1575.

Chardy, P., M. Glemarec, and A. Laurec (1976). Application of inertia methods to benthic marine ecology: practical implications of the basic options. *Estuary and Coastal Marine Science* **4**:179–205.

Christensen, N. L. (1977). Changes in structure, pattern and diversity associated with climax forest maturation in Piedmont, North Carolina. *American Midland Naturalist* **97**:176–188.

Clapham, A. R. (1936). Over-dispersion in grassland communities and the use of statistical methods in plant ecology. *Journal of Ecology* **24**:232–251.

Claridge, M. F. and M. R. Wilson (1981). Host plant associations, diversity and species-area relationships of mesophyll-feeding leafhoppers of trees and shrubs in Britain. *Environmental Entomology* **6**:217–238.

Clark, P. J. and F. C. Evans (1954). Distance to nearest neighbour as a measure of spatial relationships in populations. *Ecology* **35**:445–453.

Clifford, H. T. and W. Stephenson (1975). *An Introduction to Numerical Classification*. Academic Press, New York, pp. 37–46.

Clifford, H. T. and W. T. Williams (1976). Similarity measures. In *Pattern Analysis in Agricultural Science* (W. T. Williams, Ed.). Elsevier, New York.

Clymo, R. S. (1980). Preliminary survey of the peat-bog Hummell Knowe Moss using various numerical methods. *Vegetatio* **42**:129–148.

Cochran, W. G. (1963). *Sampling Techniques*, 2nd ed. Wiley, New York.

Cody, M. L. (1974). *Competition and the Structure of Bird Communities*. Princeton University Press, Princeton, NJ.

Coetzee, B. J. and M. J. A. Werger (1975). On association analysis and the classification of plant communities. *Vegetatio* **30**:201–206.

Connell, J. H. (1963). Territorial behaviour and dispersion in some marine invertebrates. *Research in Population Ecology* **5**:87–101.

Conner, R. N. and C. S. Adkinson (1976). Discriminate function analysis: A possible aid in determining the impact of forest management on woodpecker nesting habitat. *Forest Science* **22**:122–127.

Cook, C. W. and R. Hurst (1963). A quantitative measure of plant association in ranges in good and poor condition. *Journal of Range Management* **15**:266–274.

Cooley, W. W. and P. R. Lohnes (1971). *Multivariate Data Analysis*. Wiley, New York.

Cottam, G. and J. T. Curtis (1956). The use of distance measures in phytosociological sampling. *Ecology* 37:451–460.

Cottam, G., F. G. Goff, and R. H. Whittaker (1978). Wisconsin comparative ordination. In *Ordination of Plant Communities* (R. H. Whittaker, Ed.). W. Junk, The Hague, pp. 185–213.

Cox, G. W. (1985). *Laboratory Manual of General Ecology*, 5th ed. Brown, Dubuque, IA.

Cox, T. F. and T. Lewis (1976). A conditioned distance ratio method for analyzing spatial patterns. *Biometrika* 63:483–491.

Crawford, R. M. M. and D. Wishart (1967). A rapid multivariate method for the detection and classification of groups of ecologically related species. *Journal of Ecology* 55:505–524.

Crawford, R. M. M. and D. Wishart (1968). A rapid classification and ordination method and its application to vegetation mapping. *Journal of Ecology* 56:385–404.

Crow, T. R. and D. F. Grigal (1979). A numerical analysis of arborescent communities in the rain forest of the Luquillo Mountains, Puerto Rico. *Vegetatio* 40:135–146.

Culp, J. M. and R. W. Davies (1980). Reciprocal averaging and polar ordination as techniques for analyzing lotic macroinvertic communities. *Canadian Fishery and Aquatic Science* 37:1358–1364.

Curtis, J. T. and R. P. McIntosh (1951). The upland forest continuum in the prairie-forest border region of Wisconsin. *Ecology* 32:476–496.

Dale, M. B. and H. T. Clifford (1976). On the effectiveness of higher taxonomic ranks for vegetation analysis. *Australian Journal of Ecology* 1:37–62.

Danin, A. (1976). Plant species diversity under desert conditions I. Annual species diversity in the Dead Sea Valley. *Oecologia* 22:251–259.

Dargie, T. C. D. (1984). On the integrated interpretation of indirect site ordinations: a case study using semi-arid vegetation in southeastern Spain. *Vegetatio* 55:37–55.

David, F. N. and P. G. Moore (1954). Notes on contagious distributions in plant populations. *Annals of Botany* 18:47–53.

De Jong, T. M. (1975). A comparison of three diversity indices based on their components of richness and evenness. *Oikos* 26:222–227.

De Pablo, C. L., B. Peco, E. F. Galiano, J. P. Nicolas, and F. D. Pineda (1982). Space-time variability in mediterranean pastures analyzed with diversity parameters. *Vegetatio* 50:113–125.

Del Moral, R. (1972). Diversity patterns in forest vegetation of the Wenatchee Mountains, Washington. *Bulletin Torrey Botanical Club* 99:57–64.

Dice, L. R. (1945). Measures of the amount of ecological association between species. *Ecology* 26:297–302.

Dice, L. R. (1952). Measure of the spacing between individuals within a population. *Contribution Laboratory Vertebrate Biology*, University of Michigan, East Lansing, MI.

Diggle, P. J. (1983). *Statistical Analysis of Spatial Point Patterns*. Academic Press, New York.

Diggle, P. J., J. Besag, and J. T. Gleaves (1976). Statistical analysis of spatial point patterns by means of distance methods. *Biometrics* **32**:659–667.

Dix, R. L. and J. E. Butler (1960). A phytosociological study of a small prairie in Wisconsin. *Ecology* **41**:316–327.

Dixon, W. J. and M. B. Brown (Eds.) (1979). *BMDP-79: Biomedical Computer Programs, P-Series*. University of California Press, Berkeley, CA.

Doncaster, C. P. (1981). The spatial distribution of ants' nests on Ramsey Island, South Wales. *Journal of Animal Ecology* **50**:195–218.

Drake, J. A. (1984). Species aggregation: the influence of detritus in a benthic invertebrate community. *Hydrobiology* **112**:109–115.

Duncan, T. and G. F. Estabrook (1976). An operational method for evaluating classifications. *Systematic Botany* **1**:373–382.

Eberhardt, L. L. (1967). Some developments in distance sampling. *Biometrics* **23**:207–216.

Eberhardt, L. L. (1976). Quantitative ecology and impact assessment. *Journal of Environmental Management* **4**:27–70.

Elliott, J. M. (1973). *Some Methods for the Statistical Analysis of Samples of Benthic Invertebrates*. Scientific Publication No. 25, Freshwater Biological Association, Ambleside, Westmorland, Great Britain.

Erman, D. C. (1973). Ordination of some littoral benthic communities in Bear Lake, Utah-Idaho. *Oecologia* **13**:211–226.

Everitt, B. (1974). *Cluster Analysis*. Wiley, New York.

Faith, D. P. (1983). Asymmetric binary similarity measures. *Oecologia* **57**:287–290.

Faith, D. P. (1984). Patterns of sensitivity of association measures in numerical taxonomy. *Mathematical Biosciences* **69**:199–207.

Faith, D. P., M. P. Austin, L. Belbin, and C. R. Margules (1985). Numerical classifications of profile attributes in environmental studies. *Journal of Environmental Management* **20**:73–85.

Faith, D. P., P. R. Minchin, and L. Belbin (1987). Compositional dissimilarity as a robust measure of ecological distance. *Vegetatio* **69**:57–68.

Fasham, M. J. R. (1977). A comparison of nonmetric multidimensional scaling, principal components and reciprocal averaging for the ordination of simulated coenoclines, and coenoplanes. *Ecology* **58**:551–561.

Feoli, E. and M. Lagonegro (1983). A resemblance function based on probability: Applications to field and simulated data. *Vegetatio* **53**:3–9.

Fewster, P. H. and L. Orloci (1983). On choosing a resemblance measure for non-linear predictive ordination. *Vegetatio* **54**:27–35.

Findley, J. S. and M. T. Findley (1985). A search for pattern in butterfly fish communities. *American Naturalist* **126**:800–816.

Fish, E. B. (1976). Comparisons of phytosociological methods of classification on a desert grassland site. *Agro-Ecosystems* **2**:173–194.

Fisher, R. A. (1936). The use of multiple measurements in taxonomic problems. *Annals of Human Genetics* **7**:179–188.

Foran, B. D., G. Bastin, and K. A. Shaw (1986). Range assessment and monitoring arid lands: the use of classification and ordination in range survey. *Journal Environmental Management* **22**:67–84.

Forsythe, W. L. and O. L. Loucks (1972). A transformation for species response to habitat factors. *Ecology* **53**:1112–1119.

Garratt, M. W. and R. K. Steinhorst (1976). Testing for significance of Morisita's, Horn's and related measures of overlap. *American Midland Naturalist* **96**:245–251.

Gauch, H. G. (1973). The relationship between sample similarity and ecological distance. *Ecology* **54**:618–622.

Gauch, H. G. (1977). *ORDIFLEX – A Flexible Computer Program for Four Ordination Techniques: Weighted Averages, Polar Ordination, Principal Components Analysis, and Reciprocal Averaging.* Release B, Cornell University, Ithaca, NY.

Gauch, H. G. (1982). *Multivariate Analysis in Community Ecology.* Cambridge University Press, New York.

Gauch, H. G. and W. M. Scruggs (1979). Variants of polar ordination. *Vegetatio* **40**:147–153.

Gauch, H. G. and E. L. Stone (1979). Vegetation and soil pattern in a mesophytic forest at Ithaca, New York. *American Midland Naturalist* **102**:332–345.

Gauch, H. G. and T. R. Wentworth (1976). Canonical correlation analysis as an ordination technique. *Vegetatio* **33**:17–22.

Gauch, H. G. and R. H. Whittaker (1972). Comparison of ordination techniques. *Ecology* **53**:868–875.

Gauch, H. G., G. B. Chase, and R. H. Whittaker (1974). Ordination of vegetation samples by Gaussian species distributions. *Ecology* **55**:1382–1390.

Gauch, H. G., R. H. Whittaker, and T. R. Wentworth (1977). A comparative study or reciprocal averaging and other ordination techniques. *Journal of Ecology* **65**:157–174.

Gause, G. F. (1934). *Struggle for Existence.* Hafner, New York.

George, D. G. and R. W. Edwards. (1976). The effect of wind on the distribution of chlorophyll *a* and crustacean plankton in a shallow eutrophic reservoir. *Journal of Applied Ecology* **13**:667–690.

Gerdol, R., C. Ferrari, and F. Piccoli (1985). Correlation between soil characters and forest types: A study in multiple discriminant analysis. *Vegetatio* **60**:49–56.

Gibson, D. J. and P. Greig-Smith (1986). Community pattern analysis: A method for quantifying community mosaic structure. *Vegetatio* **66**:41–47.

Giller, P. S. (1984). *Community Structure and the Niche.* Chapman and Hall, London.

Gittens, R. (1965). Multivariate approaches to a limestone grassland. I. A stand ordination. *Journal of Ecology* **53**:385–401.

Gittens, R. (1985). *Canonical Analysis: A Review with Applications in Ecology.* Springer-Verlag, Berlin.

Gladfelter, W. B. and W. S. Johnson (1983). Feeding niche separation in a guild of tropical reef fishes (*Holocentridae*). *Ecology* **64**:552–563.

Gleason, H. A. (1926). The individualistic concept of the plant association. *Bulletin Torrey Botanical Club* **53**:7–26.

Goff, F. G. (1975). Comparison of species ordinations resulting from alternative indices of interspecific association and different numbers of included species. *Vegetatio* **31**:1–14.

Goodall, D. W. (1953). Objective methods for the classification of vegetation. I. The use of positive interspecific correlation. *Australian Journal of Botany* **1**:39–63.

Goodall, D. W. (1954a). Minimal area: a new approach. *VIIth International Botanical Congress.* Rapp. Comm. Parv. avant le Congress, Section 7, Ecologie, pp. 19–21.

Goodall, D. W. (1954b). Objective methods for the classification of vegetation. III. An essay in the use of factor analysis. *Australian Journal of Botany* **2**:304–324.

Goodall, D. W. (1963). Pattern analysis and minimal area—some further comments. *Journal of Ecology* **51**:705–710.

Goodall, D. W. (1964). A probabilistic index. *Nature, London* **203**:1098.

Goodall, D. W. (1966). A new similarity index based on probability. *Biometrics* **22**:882–907.

Goodall, D. W. (1970). Statistical plant ecology. *Annual Review of Ecology and Systematics* **1**:99–124.

Goodall, D. W. (1973). Sample similarity and species correlation. In *Ordination and Classification of Communities* (R. H. Whittaker, Ed.). W. Junk, The Hague, pp. 105–156.

Goodall, D. W. (1974). A new method for the analysis of spatial pattern by random pairing of quadrats. *Vegetatio* **29**:135–146.

Goodall, D. W. (1978a). Numerical classification. In *Classification of Plant Communities* (R. H. Whittaker, Ed.). W. Junk, The Hague, pp. 247–286.

Goodall, D. W. (1978b). Sample similarity and species correlation. In *Ordination of Plant Communities* (R. H. Whittaker Ed.). W. Junk, The Hague, pp. 99–149.

Goodall, D. W. and N. E. West (1979). A comparison of techniques for assessing dispersion patterns. *Vegetatio* **40**:15–28.

Gower, J. C. (1966). Some distance properties of latent root and vector methods used in multivariate analysis. *Biometrika* **53**:325–338.

Gray, J. S. (1981). Detecting pollution induced changes in communities using the log-normal distribution of individuals among species. *Marine Pollution Bulletin* **12**:173–176.

Green, R. H. (1966). Measurement of non-randomness in spatial distributions. *Researches Population Ecology* **8**:1–7.

Green, R. H. (1979). *Sampling Design and Statistical Methods for Environmental Biologists.* Wiley, New York.

Green, R. H. (1980). Multivariate approaches in ecology: the assessment of ecological similarity. *Annual Review of Ecology and Systematics* **11**:1–14.

Greig-Smith, P. (1952a). The use of random and contiguous quadrats in the study of the structure of plant communities. *Annals of Botany* **16**:293–316.

Greig-Smith, P. (1952b). Ecological observations on degraded and secondary forests in Trinidad, British West Indies. *Journal of Ecology* **40**:316–330.

Greig-Smith, P. (1971). Analysis of vegetation data: the user view-point. In *Statistical Ecology* (G. P. Patil, E. C. Pielou, and W. E. Waters, Eds.), Pennsylvania State University Press, University Park, PA, Volume 3, pp. 149–166.

Greig-Smith, P. (1979). Presidential address 1979: Pattern in vegetation. *Journal of Ecology* **67**:755–780.

Greig-Smith, P. (1983). *Quantitative Plant Ecology*, 3rd ed. University of California Press, Berkeley, CA.

Gulmon, S. L. and H. A. Mooney (1977). Spatial and temporal relationships between two desert shrubs, *Atriplex hymenolytra* and *Tidestromia oblongifolia* in Death Valley, California. *Journal of Ecology* **65**:831–838.

Hajdu, L. J. (1981). Graphical comparison of resemblance measures in phytosociology. *Vegetatio* **48**:47–59.

Hartigan, J. A. (1975). *Clustering Algorithms*. Wiley, New York.

Heip, C. (1974). A new index measuring evenness. *Journal of Marine Biological Association* **54**:555–557.

Heltshe, J. F. and J. DiCanzio (1985). Power study of jackknifed diversity indices to detect change. *Journal of Environmental Management* **21**:331–341.

Heltshe, J. F. and N. E. Forrester (1983a). Estimating species richness using the jackknife procedure. *Biometrics* **39**:1–11.

Heltshe, J. F. and N. E. Forrester (1983b). Estimating diversity using quadrat sampling. *Biometrics* **39**:1073–1076.

Heltshe, J. F. and N. E. Forrester (1985). Statistical evaluation of the jackknife estimate of diversity when using quadrat samples. *Ecology* **66**:107–111.

Hendrickson, J. A. (1979). The biological motivation for abundance models. In *Statistical Distributions in Ecological Work* (J. K. Ord, G. P. Patil, and C. Taillie, Eds.). International Co-operative Publ., Fairland, MD, pp. 263–274.

Hill, M. O. (1973a). The intensity of spatial pattern in plant communities. *Journal of Ecology* **61**:225–236.

Hill, M. O. (1973b). Diversity and evenness: A unifying notation and its consequences. *Ecology* **54**:427–432.

Hill, M. O. (1973c). Reciprocal averaging: An eigenvector method of ordination. *Journal of Ecology* **61**:237–249.

Hill, M. O. (1974). Correspondence analysis: A neglected multivariate method. *Journal of the Royal Statistical Society, Series C* **23**:340–354.

Hill, M. O. and H. G. Gauch (1980). Detrended correspondence analysis, an improved ordination technique. *Vegetatio* **42**:47–58.

Hill, R. S. (1980). A stopping rule for partitioning dendrograms. *Botanical Gazette* **141**:321–324.

Holgate, P. (1965). Some new tests of randomness. *Journal of Ecology* **53**:261–266.

Horn, H. (1966). Measurement of "overlap" in comparative ecological studies. *American Naturalist* **100**:419–424.

Hubalek, Z. (1982). Coefficients of association and similarity based on binary (presence-absence) data: an evaluation. *Biological Reviews* **57**:669–689.

Hubbell, S. P. (1979). Tree dispersion, abundance, and diversity in a tropical dry forest. *Science* **203**:1299–1309.

Huhta, V. (1979). Evaluation of different similarity indices as measures of succession in arthropod communities of the forest floor after clear-cutting. *Oecologia* **41**:11–23.

Hurlbert, S. H. (1969). A coefficient of interspecific association. *Ecology* **50**:1–9.

Hurlbert, S. H. (1971). The non-concept of species diversity: A critique and alternative parameters. *Ecology* **52**:577–586.

Hurlbert, S. H. (1978). The measurement of niche overlap and some relatives. *Ecology* **59**:67–77.

Hurlbert, S. H. (1982). Notes on the measurement of overlap. *Ecology* **63**:252–253.

Hurlbert, S. H. and J. O. Keith (1979). Distribution and spatial patterning of flamingos in the Andean Altiplano. *Auk* **96**:328–342.

Hutchinson, G. E. (1953). The concept of pattern in ecology. *Proceedings Academy Natural Sciences*, Philadelphia, PA.

Hutchinson, G. E. (1959). Homage to Santa Rosalia, or why are there so many kinds of animals? *American Naturalist* **93**:145–159.

Ihm, P. and H. van Groenewoud (1975). A multivariate ordering of vegetation data based on Gaussian type gradient response curves. *Journal of Ecology* **63**:767–777.

Jaccard, P. (1908). Nouvelles recherches sur la distribution florale. *Bulletin Society Sciences Naturale* **44**:223–270.

James, F. C. and S. Rathbun (1981). Rarefraction, relative abundance, and diversity of avian communities. *Auk* **98**:785–800.

Janson, S. and J. Vegelius (1981). Measures of ecological association. *Oecologia* **49**:371–376.

Jeglum, J. K., C. F. Wehrhahn, and J. M. A. Swan (1971). Comparisons of environmental ordinations with principal components—vegetational ordinations for sets of data having different complexities. *Canadian Journal of Forest Research* **1**:99–112.

Jensen, D. D., G. B. Beus, and G. Storm (1968). Simultaneous statistical tests on categorical data. *Journal of Experimental Education* **36**:46–56.

Jensen, S. (1978). Influences of transformation of cover values on classification and ordination of lake vegetation. *Vegetatio* **37**:19–31.

Jensen, S. (1979). Classification of lakes in southern Sweden on the basis of their macrophyte composition by means of multivariate methods. *Vegetatio* **39**:129–146.

Jesberger, J. A. and J. W. Sheard (1973). A quantitative study and multivariate analysis of corticolous lichen communities in the southern boreal forest of Saskatchewan. *Canadian Journal of Botany* 51:185–201.

Johnson, R. A. and D. W. Wichern (1982). *Applied Multivariate Statistical Analysis.* Prentice-Hall, Englewood Cliffs, NJ.

Johnson, R. B. and W. J. Zimmer (1985). A more powerful test for dispersion using distance measurements. *Ecology* 66:1084–1085.

Johnson, R. W. and D. W. Goodall (1979). A maximum likelihood approach to non-linear ordination. *Vegetatio* 41:133–142.

Kaesler, R. L. and E. E. Herricks (1976). Analysis of data from biological surveys of streams: diversity and sample size. *Water Resources Bulletin* 12:125–135.

Kempton, R. A. and L. R. Taylor (1978). The Q-statistic and the diversity of floras. *Nature* 275:252–253.

Kenkel, N. C. and L. Orloci (1986). Applying metric and nonmetric multidimensional scaling to ecological studies: some new results. *Ecology* 67:919–928.

Kershaw, K. A. (1961). Association and co-variance analysis of plant communities. *Journal of Ecology* 49:643–655.

Kershaw, K. A. (1973). *Quantitative and Dynamic Plant Ecology,* 2nd ed. Elsevier, New York.

Kessell, S. R. and R. H. Whittaker (1976). Comparisons of three ordination techniques. *Vegetatio* 32:21–29.

Kolasa, J. and E. Biesiadka (1984). Diversity concept in ecology. *Acta Biotheoretica* 33:145–162.

Knapp, R. (Ed.) (1984). *Sampling Methods and Taxon Analysis in Vegetation Science.* W. Junk, The Hague.

Kruskal, J. B. (1964). Nonmetric multidimensional scaling: A numerical method. *Psychometrika* 29:115–129.

Lamacraft, R. R., M. H. Friedel, and V. H. Chewings (1983). Comparison of distance based density estimates for some arid rangeland vegetation. *Australian Journal of Ecology* 8:181–187.

Lambert, J. M. and W. T. Williams (1962). Multivariate methods in plant ecology. IV. Nodal analysis. *Journal of Ecology* 50:775–802.

Lamont, B. B. and J. E. D. Fox (1981). Spatial pattern of six sympatric leaf variants and two size classes of *Acacia aneura* in a semi-arid region of Western Australia. *Oikos* 37:73–79.

Lamont, B. B., S. Downs, and J. E. D. Fox (1977). Importance-value curves and diversity indices applied to a species-rich heathland in Western Australia. *Nature* 265:438–441.

Lance, G. N. and W. T. Williams (1967). A general theory for classificatory sorting strategies. 1. Hierarchical systems. *Computer Journal* 9:373–380.

Lance, G. N. and W. T. Williams (1968). A general theory for classificatory sorting strategies. 2. Clustering systems. *Computer Journal* 10:271–276.

Lawlor, L. P. (1980). Overlap, similarity and competition coefficients. *Ecology* **61**:245–251.

Lawson, G. W. (1978). The distribution of seaweed floras in the tropical and subtropical Atlantic Ocean: A quantitative approach. *Botanical Journal of the Linnean Society* **76**:177–197.

Lehman, J. T. and D. Scavia (1982). Microscale patchiness of nutrients in plankton communities. *Science* **216**:729–730.

Legendre, L. and P. Legendre (1983). *Numerical Ecology*. Elsevier, New York.

Legendre, P., S. Dallot, and L. Legendre (1985). Succession of species within a community: Chronological clustering, with applications to marine and freshwater zooplankton. *American Naturalist* **125**:257–288.

Levins, R. (1968). *Evolution in Changing Environments: Some Theoretical Explorations*. Princeton University Press, Princeton, NJ.

Lewis, T. G. and J. W. Doerr (1976). *Minicomputers: Structure and Programming*. Hayden Book, Rochelle Park, NJ.

Livingston, R. J. (1976). Diurnal and seasonal fluctuations of organisms in a north Florida estuary. *Estuarine and Coastal Marine Science* **4**:373–400.

Livingston, R. J., G. J. Kobylinski, F. G. Lewis, and P. F. Sheridan (1976). Long-term fluctuations of epibenthic fish and invertebrate populations in Apalachicole Bay, Florida. *Fishery Bulletin* **74**:311–321.

Lloyd, M. (1967). Mean crowding. *Journal of Animal Ecology* **36**:1–30.

Ludwig, J. A. (1979). A test of different quadrat variance methods for the analysis of spatial pattern. In *Spatial and Temporal Analysis in Ecology*, (R. M. Cormack and J. K. Ord, Eds.). International Co-operative Publ., Fairland, MD, pp. 289–304.

Ludwig, J. A. and D. W. Goodall (1978). A comparison of paired- with blocked-quadrat variance methods for the analysis of spatial pattern. *Vegetatio* **38**:49–59.

Lyons, N. I. (1981). Comparing diversity indices based on counts weighted by biomass or other importance values. *American Naturalist* **118**:438–442.

Madgwick, H. A. I. and P. A. Desrochers (1972). Association-analysis and the classification of forest vegetation of the Jefferson National Forest. *Journal of Ecology* **60**:285–292.

Mahalanobis, P. C. (1936). On the generalized distance in statistics. *Proceedings of the National Institute of Science of India* **12**:49–55.

Majer, J. D. (1976). The influence of ants and ant manipulation on the cocoa farm fauna. *Journal of Applied Ecology* **13**:157–175.

Margalef, R. (1958). Information theory in ecology. *General Systematics* **3**:36–71.

Martinka, R. R. (1972). Structural characteristics of blue grouse territories in southeastern Montana. *Journal of Wildlife Management* **36**:498–510.

Mason, C. F. (1977). The performance of a diversity index in describing the zoobenthos of two lakes. *Journal of Applied Ecology* **14**:363–367.

Matta, J. F. and H. G. Marshall (1984). A multivariate analysis of phytoplankton assemblages in the western North Atlantic. *Journal of Plankton Research* **6**:663–675.

Matthews, J. A. (1978). An application of non-metric multidimensional scaling to the construction of an improved species plexus. *Journal of Ecology* **66**:157–173.

May, R. M. (1975). Pattern of species abundance and diversity. In *Ecology and Evolution of Communities* (M. L. Cody and J. M. Diamond, Eds.). Belnap Press, Cambridge, MA, pp. 81–120.

May, R. M. (1981). Patterns in multi-species communities. In *Theoretical Ecology* (R. M. May, Ed.). Sinauer Assoc., Sunderland, MA, pp. 197–227.

McCulloch, C. E. (1985). Variance tests for species association. *Ecology* **66**:1676–1681.

McDonald, R. P. (1962). A general approach to nonlinear factor analysis. *Psychometrika* **27**:397–415.

McDonald, R. P. (1967). Numerical methods for polynomial models in nonlinear factor analysis. *Psychometrika* **32**:77–112.

McGuinness, K. A. (1984). Equations and explanations in the study of species-area curves. *Biological Reviews* **59**:423–440.

McIntosh, R. P. (1967). The continuum concept of vegetation. *Botanical Reviews* **33**:130–187.

McIntosh, R. P. (1978). Matrix and plexus techniques. In *Ordination of Plant Communities* (R. H. Whittaker, Ed.). W. Junk, The Hague, pp. 151–184.

McNaughton, S. J. and L. L. Wolf (1970). Dominance and the niche in ecological systems. *Science* **167**:131–139.

Menhinick, E. F. (1964). A comparison of some species-individuals diversity indices applied to samples of field insects. *Ecology* **45**:859–861.

Minchin, P. R. (1987). An evaluation of the relative robustness of techniques for ecological ordination. *Vegetatio* **69**:89–107.

Minshall, G. W., R. C. Petersen, Jr., and C. F. Nimz (1985). Species richness in streams of different size from the same drainage basin. *American Naturalist.* **125**:16–38.

Monk, C. D. (1967). Tree species diversity in the eastern deciduous forest with particular reference to north central Florida. *American Naturalist* **101**:173–187.

Morisita, M. (1959). Measuring of interspecific association and similarity between communities. *Memoirs Faculty Kyushu University, Series E* **3**:65–80.

Morisita, M. (1971). Composition of the I_d index. *Researches in Population Ecology* **13**:1–27.

Mueller, L. D. and L. Altenberg (1985). Statistical inferences on measures of niche overlap. *Ecology* **66**:1204–1210.

Mueller-Dombois, D. (1974). *Aims and Methods of Vegetation Ecology.* Wiley, New York.

Myers, J. H. (1978). Selecting a measure of dispersion. *Environmental Entomology* **7**:619–621.

Myers, W. L. and R. L. Shelton (1980). *Survey Methods for Ecosystem Management.* Wiley, New York.

Nash, C. B. (1950). Association between fish species in tributaries and shore waters of western Lake Erie. *Ecology* 31:561–566.

Nelder, J. A. (1975). *Computers in Biology*. Wykeham Publishing, London.

Newsome, R. D. and R. L. Dix (1968). The forests of the Cypress Hills, Alberta and Saskatchewan, Canada. *American Midland Naturalist* 80:118–185.

Niering, W. A. and C. H. Lowe (1984). Vegetation of the Santa Catalina Mountains: community types and dynamics. *Vegetatio* 58:3–28.

Norris, J. M. and J. P. Barkham (1970). A comparison of some Cotswold beechwoods using multiple-discriminant analysis. *Journal of Ecology* 58:603–619.

Noy-Meir, I. (1970). *Component analysis of semi-arid vegetation in southeastern Australia*. Ph.D. Dissertation, Australian National University, Canberra.

Noy-Meir, I. (1971). Multivariate analysis of the semi-arid vegetation in southeastern Australia: Nodal ordination by component analysis. *Proceedings of the Ecological Society of Australia*. 6:159–193.

Noy-Meir, I. (1973). Data transformations in ecological ordinations. I. Some advantages of non-centering. *Journal of Ecology* 61:329–341.

Noy-Meir, I. (1974). Catenation: Quantitative methods for the definition of coenoclines. *Vegetatio* 29:89–99.

Noy-Meir, I. and M. P. Austin (1970). Principal component ordination and simulated vegetation data. *Ecology* 51:193–215.

Noy-Meir, I. and R. H. Whittaker (1977). Continuous multivariate methods in community analysis: Some problems and developments. *Vegetatio* 33:79–98.

Noy-Meir, I. and R. H. Whittaker (1978). Recent developments in continuous multivariate techniques. In *Ordination of Plant Communities* (R. H. Whittaker, Ed.). W. Junk, The Hague, pp. 337–378.

Noy-Meir, I., D. Walker, and W. T. Williams (1975). Data transformations in ecological ordinations. II. On the meaning of data standardization. *Journal of Ecology* 63: 779–800.

Ochiai, A. (1957). Zoogeographic studies on the soleoid fishes found in Japan and its neighbouring regions. *Bulletin Japanese Soc. Sci. Fisheries* 22:526–530.

Oksanen, J. (1983). Ordination of boreal heath-like vegetation with principal component analysis, correspondence analysis, and multidimensional scaling. *Vegetatio* 52:181–189.

Orloci, L. (1966). Geometric models in ecology. I. The theory and application of some ordination methods. *Journal of Ecology* 54:193–215.

Orloci, L. (1967a). An agglomerative method for classification of plant communities. *Journal of Ecology* 55:193–206.

Orloci, L. (1967b). Data centering: A review and evaluation with reference to component analysis. *Systematic Zoology* 16:208–212.

Orloci, L. (1972). On objective functions of phytosociological resemblance. *American Midland Naturalist* 88:28–55.

Orloci, L. (1973). Ordination by resemblance matrices. In *Ordination and Classification*

(R. H. Whittaker, Ed.). W. Junk, The Hague, pp. 251–286.

Orloci, L. (1974a). On information flow in ordination. *Vegetatio* **29**:11–16.

Orloci, L. (1974b). Revisions for the Bray and Curtis ordination. *Canadian Journal of Botany* **52**:1773–1776.

Orloci, L. (1978). *Multivariate Analysis in Vegetation Research*, 2nd ed. W. Junk, The Hague.

Orloci, L. (1980). An algorithm for predictive ordination. *Vegetatio* **42**:23–26.

Orloci, L. and N. C. Kenkel (1985). *Introduction to Data Analysis with Examples from Population and Community Ecology*. Statistical Ecology Monographs, Volume 1, International Co-operative Publishing House, Fairland, MD.

Pearson, K. (1901). On lines and planes of closest fit to systems of points in space. *Philosophical Magazine, Sixth Series* **2**:559–572.

Peet, R. K. (1974). The measurement of species diversity. *Annual Review of Ecology and Systematics* **5**:285–307.

Peet, R. K. (1975). Relative diversity indices. *Ecology* **56**:496–498.

Peet, R. K. and O. L. Loucks (1977). A gradient analysis of southern Wisconsin forests. *Ecology* **58**:485–499.

Pemberton, S. G. and R. W. Frey (1984). Quantitative methods in ichnology: spatial distribution among populations. *Lethaia* **17**:33–49.

Petraitis, P. S. (1979). Likelihood measures of niche breadth and overlap. *Ecology* **60**:703–710.

Petraitis, P. S. (1985). The relationship between likelihood niche measures and replicated tests for goodness of fit. *Ecology* **66**:1983–1985.

Phillips, D. L. (1978). Polynomial ordination: Field and computer simulation testing of a new method. *Vegetatio* **37**:129–140.

Pianka, E. R. (1973). The structure of lizard communities. *Annual Review of Ecology and Systematics* **4**:53–74.

Pielou, E. C. (1959). The use of point-to-plant distances in the study of the pattern of plant populations. *Journal of Ecology* **47**:607–615.

Pielou, E. C. (1966). The measurement of diversity in different types of biological collections. *Journal of Theoretical Biology* **13**:131–144.

Pielou, E. C. (1972a). Niche width and overlap: a method for measuring them. *Ecology* **53**:687–692.

Pielou, E. C. (1972b). 2^k contingency tables in ecology. *Journal of Theoretical Biology* **34**:337–352.

Pielou, E. C. (1974). *Population and Community Ecology*. Gordon and Breach, New York.

Pielou, E. C. (1975). *Ecological Diversity*. Wiley, New York.

Pielou, E. C. (1977). *Mathematical Ecology*. Wiley, New York.

Pielou, E. C. (1979). *Biogeography*. Wiley, New York.

Pielou, E. C. (1984). *The Interpretation of Ecological Data*. Wiley, New York.

Pimental, R. A. (1979). *Morphometrics.* Kendall/Hunt, Dubuque, IA.

Pimm, S. L. and D. P. Bartell (1980). Statistical model for predicting range expansion of the red imported fire ant, *Solenopsis invicta*, in Texas. *Environmental Entomology* 9:653–658.

Poole, R. W. (1974). *An Introduction to Quantitative Ecology.* McGraw-Hill, New York.

Popma, J., L. Mucina, O. van Tongeren, and E. van der Maarel (1983). On the determination of optimal levels in phytosociological classifications. *Vegetatio* 52:65–75.

Prentice, I. C. (1977). Non-metric ordination methods in ecology. *Journal of Ecology* 65:85–94.

Preston, F. W. (1948). The commonness, and rarity, of species. *Ecology* 29:254–283.

Preston, F. W. (1962). The canonical distribution of commonness and rarity, Part I. *Ecology* 43:185–215.

Pyke, G. H. (1982). Local geographic distributions of bumblebees near Crested Butte, Colorado: Competition and community structure. *Ecology* 63:555–573.

Quinn, J. F. and A. E. Dunham (1983). On hypothesis testing in ecology and evolution. *American Naturalist* 122:602–617.

Ratliff, R. D. (1982). A correction of Cole's C7 and Hurlbert's C8 coefficients of inter-specific association. *Ecology* 63:1605–1606.

Ratliff, R. D. and R. D. Pieper (1981). Deciding final clusters: an approach using intra- and intercluster distances. *Vegetatio* 48:83–86.

Ratliff, R. D. and R. D. Pieper (1982). Approaches to plant community classification for range managers. *Journal of Range Management Monograph Series*, No. 1.

Ray, A. A. (Ed.) (1982). *SAS User's Guide: Statistics*, 1982 Edition. SAS Institute, Inc., Cary, NC.

Revelante, N. and M. Gilmartin (1980). Microplankton diversity indices as indicators of eutrophication in the Northern Adriatic Sea. *Hydrobiolgia* 70:277–286.

Rice, J., R. D. Ohmart, and B. W. Anderson (1983). Habitat selection attributes of an avian community: A discriminant analysis investigation. *Ecological Monographs* 53:263–290.

Ricklefs, R. E. and M. Lau (1980). Bias and dispersion of overlap indices: Results of some Monte Carlo simulations. *Ecology* 61:1019–1024.

Roberts, D. W. (1986). Ordination on the basis of fuzzy set theory. *Vegetatio* 66:123–131.

Rohlf, F. J. (1974). Methods of comparing classifications. *Annual Review of Ecology and Systematics* 5:101–113.

Rohlf, F. J. (1982). Consensus indices for comparing classifications. *Mathematical Biological Sciences* 59:131–144.

Rohlf, F. J. and R. R. Sokal (1981). *Statistical Tables.* Freeman, San Francisco, CA.

Rohlf, F. J., J. Kishpaugh, and D. Kirk (1971). *NT-SYS Numerical Taxonomy System of Multivariate Statistical Programs.* Technical Report, State University of New York at Stony Brook, Stony Brook, NY.

Romesburg, H. C. (1984). *Cluster Analysis for Researchers*. Lifetime Learning Publications, Belmont, CA.

Root, R. B. (1967). The niche exploitation pattern of the blue-gray gnatcatcher. *Ecological Monographs* 37:317–350.

Routledge, R. D. (1979). Diversity indices: which ones are admissible? *Journal of Theoretical Biology* 76:503–515.

Sanders, H. L. (1968). Marine benthic diversity: a comparative study. *American Naturalist* 102:243–282.

Schluter, D. (1984). A variance test for detecting species associations, with some example applications. *Ecology* 65:998–1005.

Schoener, T. W. (1974). Resource partitioning in ecological communities. *Science* 185:27–39.

Scott, D. and W. D. Billings (1964). Effects of environmental factors on the standing crop and productivity of an alpine tundra. *Ecological Monographs* 34:243–270.

Searle, S. R. (1966). *Matrix Algebra for the Biological Sciences*. Wiley, New York.

Shannon, C. E. and W. Weaver (1949). *The Mathematical Theory of Communication*. University Illinois Press, Urbana, IL.

Sheldon, A. L. (1969). Equitability indices: dependence on the species count. *Ecology* 50:466–467.

Shepard R. N. (1980). Multidimensional scaling, tree-fitting, and clustering. *Science* 210:390–398.

Shepard, R. N. and J. D. Carroll (1966). Parametric representation of non-linear data structures. In *Multivariate Analysis* (P. R. Krishnaiah, Ed.). Academic Press, New York.

Simpson, E. H. (1949). Measurement of diversity. *Nature* 163:688.

Sinclair, D. F. (1985). On tests of spatial randomness using mean nearest neighbor distance. *Ecology* 66:1084–1085.

Smith, B. E. and G. Cottam. (1967). Spatial relationships of mesic herbs in southern Wisconsin. *Ecology* 48:546–558.

Smith, E. P. (1984). A note on the general likelihood measure of overlap. *Ecology* 65:323–324.

Sneath, P. H. A. and R. R. Sokal (1973). *Numerical Taxonomy*. Freeman, San Francisco, CA.

Snedecor, G. W. and W. G. Cochran (1973). *Statistical Methods*. 6th ed. Iowa State University Press, Ames, IA.

Sokal, R. R. (1974) Classification: purposes, principles, progress, prospects. *Science* 185:1115–1123.

Sokal, R. R. and F. J. Rohlf (1981). *Biometry*, 2nd ed. Freeman, San Francisco, CA.

Solomon, D. (1979). On a paradigm for mathematical modeling. In *Contemporary Quantitative Ecology and Related Ecometrics* (G. P. Patil and M. L. Rosenzweig, Eds.). International Co-operative Publishing, Fairland, MD.

Southwood, T. R. E. (1978). *Ecological Methods: With Particular Reference to the Study of Insect Populations*. Wiley, New York.

Spomer, G. G. (1973). The concepts of "interaction" and "operational environment" in environmental analyses. *Ecology* **54**:200–204.

Squires, E. R. and J. E. Klosterman (1981). Spatial patterning and competition in an aspen-white pine successional system. *American Journal of Botany* **68**:790–794.

Steele, R. G. D. and J. H. Torrie (1960). *Principles and Procedures of Statistics*. McGraw-Hill, New York.

Stephenson, W., W. T. Williams, and S. D. Cook (1972). Computer analyses of Peterson's original data on bottom communities. *Ecological Monographs* **42**:387–415.

Stephenson, W., W. T. Williams, and G. N. Lance (1970). The macrobenthos of Moreton Bay. *Ecological Monographs* **40**:459–494.

Sugihara, G. (1980). Minimal community structure: An explanation of species abundance patterns. *American Naturalist* **116**:770–787.

Swan, J. M. A. and R. L. Dix. (1966). The phytosociological structure of upland forest at Candle Lake, Saskatchewan. *Journal of Ecology* **54**:13–40.

Swanston, D. N., W. R. Meehan, and J. A. McNutt (1977). A quantitative geomorphic approach to predicting productivity of pink and chum salmon streams in southeast Alaska. *USDA Forest Service Research Paper* PNW-227, Portland, OR.

Ter Braak, C. J. F. (1986). Canonical correspondence analysis: a new eigenvector technique for multivariate direct gradient analysis. *Ecology* **67**:1167–1179.

Ter Braak, C. J. F. (1987). The analysis of vegetation-environment relationships by canonical correspondence analysis. *Vegetatio* **69**:69–77.

Thompson, D. C. (1980). A classification of the vegetation of Boothia Peninsula and the Northern District of Keewatin, N.W.T. *Arctic* **33**:73–99.

Turkington, R. and P. B. Cavers (1979). Neighbour relationships in grass-legume communities. III. Development of pattern and association in artificial communities. *Canadian Journal of Botany* **57**:2704–2710.

Usher, M. B. (1975). Analysis of pattern in real and artificial plant populations. *Journal of Ecology* **63**:569–586.

Usher, M. B. (1983). Pattern in the simple moss-turf communities of the sub-Antarctic and maritime Antarctic. *Journal of Ecology* **71**:945–958.

Van Belle, G. and I. Ahmad (1974). Measuring affinity of distributions. In *Reliability and Biometrics* (F. Proschan and R. J. Serfling, Eds.). SIAM Publications, Philadelphia, PA, pp. 651–668.

Vandermeer, J. (1981). *Elementary Mathematical Ecology*. Wiley, New York.

Vilks, G., E. H. Anthony, and W. T. Williams (1970). Application of association-analysis to distribution studies of recent *Foraminifera*. *Canadian Journal Earth Sciences* **7**:1462–1469.

Walker, B. H. and C. F. Wehrhahn (1971). Relationships between derived vegetation gradients and measured environmental variables in Saskatchewan wetlands. *Ecology* **52**:85–95.

Walker, H. A., S. B. Saila, and E. L. Anderson (1979). Exploring data structure of New York Bight benthic data using post-collection stratification of samples, and linear discriminant analysis for species composition comparisons. *Estuarine and Coastal Science* **9**:101–120.

Waring, R. H. and J. Major (1964). Some vegetation of the California coastal redwood region in relation to gradients of moisture, nutrients, light and temperature. *Ecological Monographs* **34**:167–215.

Went, F. W. (1942). The dependence of certain annual plants on shrubs in southern California deserts. *Bulletin Torrey Botanical Club* **69**:100–114.

Whittaker, R. H. (1952). A study of summer foliage insect communities in the Great Smoky Mountains. *Ecological Monographs* **22**:1–44.

Whittaker, R. H. (1965). Dominance and diversity in land plant communities. *Science* **147**:250–260.

Whittaker, R. H. (1967). Gradient analysis of vegetation. *Biological Reviews* **42**:207–264.

Whittaker, R. H. (1970). The population structure of vegetation. In *Gesellschafts-morpholgie* (R. Tuxen, Ed.). W. Junk, The Hague, pp. 39–59.

Whittaker, R. H. (1972). Evolution and measurement of species diversity. *Taxon* **21**:213–251.

Whittaker, R. H. (Ed.) (1978a). *Ordination of Plant Communities.* W. Junk, The Hague.

Whittaker, R. H. (Ed.) (1978b). *Classification of Plant Communities.* W. Junk, The Hague.

Whittaker, R. H. and H. G. Gauch (1973). Evaluation of ordination techniques. In *Ordination and Classification of Vegetation* (R. H. Whittaker, Ed.). W. Junk, The Hague, pp. 288–321.

Whittaker, R. H. and W. A. Niering (1964). Vegetation of the Santa Catalina Mountains, Arizona. I. Ecological classification and distribution of species. *Journal of the Arizona Academy of Science* **3**:9–34.

Whittaker, R. J. (1987). An application of detrended correspondence analysis and non-metric multidimensional scaling to the identification and analysis of environmental factor complexes and vegetation structures. *Journal of Ecology* **75**:363–376.

Wiegleb, G. (1981). Application of multiple discriminant analysis on the analysis of the correlation between macrophyte vegetation and water quality in running waters of Central Europe. *Hydrobiologia* **79**:91–100.

Williams, B. K. (1983). Some observations on the use of discriminant analysis in ecology. *Ecology* **64**:1283–1291.

Williams, W. T. (Ed.) (1976). *Pattern Analysis in Agricultural Science.* Elsevier, New York.

Williams, W. T. and M. B. Dale (1965). Fundamental problems in numerical taxonomy. *Advances in Botanical Research* **2**:35–68.

Williams, W. T. and J. M. Lambert (1959). Multivariate methods in plant ecology. I. Association analysis in plant communities. *Journal of Ecology* **47**:83–101.

William, W. T. and J. M. Lambert. (1961). Multivariate methods in plant ecology. III. Inverse association analysis. *Journal of Ecology* **49**:717–729.

Wishart, D. (1969). FORTRAN II programs for 8 methods of cluster analysis (Clustan I). *Kansas Geological Survey Computer Contribution*, No. 38, Kansas City, KS.

Wolda, H. (1981). Similarity indices, sample size, and diversity. *Oecologia* **50**:296–302.

Wolda, H., F. W. Fish, and M. Estribi (1983). Faunistics of Panamanian cockroaches (*Blattaria*). *Uttar Pradesh Journal of Zoology* **3**:1–9.

Yarrington, G. A. and W. G. E. Green (1966). The distributional pattern of crustose lichens on limestone cliffs at Rattlesnake Point, Ontario. *The Bryologist* **69**:450–461.

Yapp, W. B. (1979). Specific diversity in woodland birds. *Field Studies* **5**:45–58.

Yule, G. U. (1912). On the methods of measuring association in attributes. *Journal of the Royal Statistical Society, London* **75**:579–642.

Zahl, S. (1977). Jackknifing an index of diversity. *Ecology* **58**:907–913.

Zar, J. H. (1974). *Biostatistical Analysis*. Prentice-Hall, Englewood Cliffs, NJ.

Zaret, T. M. and A. S. Rand (1971). Competition in tropical stream fishes: support for the competitive exclusion principle. *Ecology* **52**:336–342.

Zaret, T. M. and E. P. Smith (1984). On measuring niches and not measuring them. In *Evolutionary Ecology of Neotropical Freshwater Fishes* (T. M. Zaret, Ed.). W. Junk, The Hague, pp. 127–137.

Index

Aggregation, *see* Spatial patterns, aggregated/clumped

Aggregation, Hopkins index of, 62–63

Association analysis, 161, 163–164, 181–188, 281, 289
nodal analysis, 187
normal *vs.* inverse, 186–187
stopping rules, 183–184, 188

Association between species, *see* Interspecific association tests

BASIC microcomputer programs, 7
community classification, 175–176, 185–187, 195–198
community interpretation, 289–290, 305
community ordination, 214, 217–220, 237, 251, 263, 267, 269
spatial pattern analysis, 32–33, 35–37, 50–51, 60–61
species abundance relations, 77, 79, 81–82, 88, 98–99
species affinity, 120–122, 138, 141, 152, 184

Bias, 8, 62–63, 87, 91–92, 95, 129, 131, 138, 140, 184–186

Bootstrap method, 123

Bray–Curtis ordination, 206–207, 212, 214, 217, 221. *See also* Polar ordination

Canonical correlation analysis, 297, 307–308

Canonical correspondence analysis, 308–309

Canonical variate analysis, 293, 295. *See also* Discriminant analysis, multiple

Central limit theorem, 74

Chi-square:
Fisher's exact test, 187
supercritical, 132
Yates' correction, 129, 135–136, 138, 140, 187

Chi-square test of goodness of fit:
lognormal distribution, 77–81
negative binomial distribution, 26, 34–40
Poisson distribution, 24, 31–32, 36–40

Chi-square test of significance:
general overlap, 117, 120
index of dispersion, 27–28, 35–39
species association, 128–129, 133–144, 153, 181–188
specific overlap, 116, 119–120

Classification, types of:
agglomerative, 160–161, 164, 187, 189
divisive, 160–161, 164, 181–182, 187
hierarchical, 160, 164, 181–183, 189, 191, 200–201
monothetic, 160–161, 164, 182, 187
polythetic, 160–161, 164, 187
reticulate, 160

Classification of communities, 5, 157, 159–202, 206, 212, 222, 277–278, 293–294

Diskette with the book: **Statistical Ecology. A Primer on Methods and Computing**
by John A. Ludwig and James F. Reynolds

o All programs on this diskette are written in BASIC for IBM PCs and compatables and are stored in ASCII format.

o Make a back-up copy of this diskette immediately and store the original. John Wiley and Company will only send a replacement if the diskette is defective.

o The authors and publisher do not assume any responsibility for any errors, mistakes, or misrepresentations that may result from the use of these programs.

o The diskette contains the following programs:

POISSON.BAS = Chi-square goodness-of-fit test for agreement with a poisson distribution

NEGBINOM.BAS = Chi-square goodness-of-fit test for agreement with a negative binominal distribution

BQV.BAS = Spatial pattern analysis by blocked-quadrat variances

PQV.BAS = Spatial pattern analysis by all possible paired-quadrat variances

TSQUARE.BAS = Spatial pattern analysis based on T-square sampling distance data

LOGNORM.BAS = Fitting observed abundance data to a lognormal distribution

RAREFRAC.BAS = Rarefraction method for species richness

SPDIVERS.BAS = Richness, diversity, and evenness indices

SPOVRLAP.BAS = General and specific indices of niche overlap

SPASSOC.BAS = Variance ratio, chi-square test, and indices for species association

SPCOVAR.BAS = R-mode correlation coefficients for species affinity

SUDIST.BAS = Q-mode distance indices for SU resemblance

NASSOC.BAS = Normal association analysis for classification of SUs

CLUSTER.BAS = Cluster analysis for classification of SUs

PO.BAS = Polar ordination

PCA.BAS = Principal component analysis ordination

COA.BAS = Correspondence analysis ordination

DPC.BAS = Detrended principal components for nonlinear ordination

NMDS.BAS = Nonmetric multidimensional scaling for nonlinear ordination

SDA.BAS = Simple discriminate analysis of classification results

PCREG.BAS = Simple and multiple regression analysis of ordination results